UK Association for Legal and Social Philosophy

The UK Association for Legal and Social Philosophy (ALSP) is open to everyone interested in the interaction between theory and practice in the areas of legal, social and political thought; the interplay between these areas; and the applications and outcomes arising out of the inter-disciplinary nature of such debate. In seeking to extend debate beyond the traditional academy, it is especially concerned to include students and practitioners in its activities, as well as to promote discussion with, among and beyond full-time academics.

Membership is currently £20 (waged) or £10 (unwaged) per annum, which includes a subscription to *Res Publica: A Journal of Social and Legal Philosophy* and the Association's newsletter. For details, please write to the Treasurer, Ms P FitzGerald, 7 Adur Court, Stoney Lane, Shoreham-by-Sea, West Sussex BN43 6LY (e-mail: fitzgerald@enterprise.net).

Justice, Property and the Environment

Social and Legal Perspectives

Edited by
TIM HAYWARD
University of Edinburgh

JOHN O'NEILL
Lancaster University

Routledge
Taylor & Francis Group

LONDON AND NEW YORK

First published 1997 by Ashgate Publishing

Reissued 2018 by Routledge
2 Park Square, Milton Park, Abingdon, Oxon, OX14 4RN
711 Third Avenue, New York, NY 10017, USA

Routledge is an imprint of the Taylor & Francis Group, an informa business

Publisher's Note
The publisher has gone to great lengths to ensure the quality of this reprint but points out that some imperfections in the original copies may be apparent.

Disclaimer
The publisher has made every effort to trace copyright holders and welcomes correspondence from those they have been unable to contact.

A Library of Congress record exists under LC control number: 97074449

ISBN 13: 978-1-138-32276-9 (hbk)
ISBN 13: 978-1-138-32277-6 (pbk)
ISBN 13: 978-0-429-45180-5 (ebk)

Contents

Acknowledgements

We would like to thank Pat FitzGerald, Bob Brecher, Liz Kingdom and Sarah Markham for all their support and assistance in producing this volume. An earlier version of 'The Merchandising of Biodiversity' by Joan Martinez-Alier was published in *Capitalism, Nature and Socialism* and is published with the permission of Guildford Press.

Introduction: social and legal perspectives on environmental problems

Tim Hayward

Environmental problems are also social problems, both in their causes and their effects. Environmental degradation is being brought about by socially organised practices, and it has generally been allowed to occur because those with the power of decision have deemed it a necessary and acceptable sacrifice justified by the benefits the practices bring. Yet there is now growing concern about whether the benefits really do always justify the damage incurred: the question thus arises of whether a different set of normative priorities may be required as part of a solution to environmental problems. As this question receives increasing attention in international fora, however, a further dimension to it also becomes increasingly evident, namely: who benefits, and who bears the losses? The *effects* of environmental degradation are not necessarily experienced as costs by the people who cause – and most benefit – from them. Indeed, many of the most serious environmental problems are not evenly distributed but are felt most acutely by people who are also already subject to socioeconomic disadvantage. Thus, as is now widely recognised, and as the cleavage between rich and poor nations at the Rio Summit, for instance, testifies, questions of environmental policy are inextricably linked with questions of social justice.

As well as the question of an equitable distribution of environmental costs and benefits among living humans, questions of justice arise in relation to two other categories: do *future generations* of humans have rights to have their needs taken into account in present decision-making about the use and distribution of natural resources? Are there some *non-human natural beings* who should be considered not simply as resources, but as beings who themselves are entitled to just treatment?

In addition to these questions of normative principle, there are also questions concerning the most appropriate and effective policies for implementing social and environmental justice. There seems to be some potential for developing

1

existing legal instruments to this end. For instance, laws of public and private liability can be so developed that stricter requirements are placed on those who cause environmental harms and greater possibilities for action are granted to those affected by them. There is also the possibility that the assignment of new environmental rights, or rights connected with genetic resources, for instance, help redress inequities in the distribution of resources.

Moreover, both the theory and application of justice in relation to the environment raise questions concerning property rights: on one view, the assignment of property rights in environmental goods or natural phenomena is a solution to the problem of the lack of incentive on identifiable parties to protect or conserve them; on another view, though, it is the interests of private-property owners themselves which most strongly conflict with environmental imperatives. Debate about this relates to the question of whether environmental harm occurs as a result of market failures, or is in fact due to the lack of regulation of market forces.

This is the range of complex and interrelated questions concerning justice, property and the environment which is explored in the present volume.

Justice and sustainability: equity, futurity and environment

Since the publication ten years ago of the influential Brundtland report, the three aims of securing environmental protection, equity in distribution and justice for future generations have been explicitly linked together in the idea of sustainability. What sustainability means in practice, though, is a matter of some considerable debate, and different views reflect different theoretical understandings. The questions that arise at the level of normative theory are investigated by a number of contributors to this volume.

Ted Benton examines the linkages between the first two general aims of sustainability, focusing on the question of how well traditional conceptions of social justice are able to accommodate the new environmental concerns. He sets out the central issues involved by identifying some key features of green thought that represent a challenge to traditional conceptions of justice. One is the idea of natural limits: this renders problematic the usual assumption in theories of justice of merely *moderate* scarcity and so also calls into question principles of distributive justice which, like Rawls's difference principle, depend on indefinite economic growth. Benton points out that under conditions of more severe scarcity, such principles have radically redistributive implications.

Another idea, based on biological, evolutionary and ethological evidence, is that similarities between humans and other animals are no less morally significant than the differences between them. Benton's 'human-animal continuism' thesis challenges the exclusive dichotomy between persons and things and provides reason to think that justice ought to be extended to at least some animals. Nevertheless, he acknowledges difficulties of theory and practice in extending justice to animals, connected to the requirement of arriving at a non-anthropocentric understanding of their goods.

Non-anthropocentric values more generally constitute the third idea whose implications for justice Benton considers. He points out that the ecocentric alternative to anthropocentrism, in particular, is problematic: as a form of moral holism which focuses on the value of systems properties it is opposed to the moralities of rights and liberty which focus on individuals. He nevertheless believes that what is true and important in ecocentric value theory is that in engaging practically and intellectually with the world we encounter its autonomy vis à vis human purposes and its awesome transcendence of our cognitive powers. This attitude of awe and wonder contrasts with instrumental valuations of nature as a bundle of resources for human use. In presenting this inspiring vision, he also shows how 'ecocentric' and 'anthropocentric' views are not simply or starkly opposed. Thus in his concluding summary of the features of a society governed by principles of green justice Benton emphasises that the green challenge does not obviate or displace traditional questions of justice, but adds new complexity to them. For instance, if appreciation of nature in itself were seen as a primary social good, akin to liberty of conscience, and if real opportunity to enjoy it were to be secured, the implications regarding distributive justice would be very great.

The quest for ecological justice is subjected to further scrutiny by Kate Soper. Like Benton, she sees the logic of any demand for ecological justice as pointing in a socialist direction. She particularly emphasises this in relation to the third dimension of sustainability, justice for future generations. Whatever obligations there may be towards future generations, she points out, cannot be obligations of humanity at large since very many humans are already deprived of the resources whose supposed availability grounds this obligation. For the 'human species' to be obliged to the future, there must also be an obligation to all those of its present number who are denied the means of survival, let alone of flourishing. This means that at the present time responsibility has to fall essentially on those sectors of the global community which have hitherto been most irresponsible and profligate in their use of global resources.

Nevertheless, if there are to be differential obligations, she believes, they must have a universally valid justification. A central argument is that any aspiration to ecological justice commits us to some theory of human need. A just resolution to the ecological crisis would require dramatic shifts in the pattern of First World consumption: to justify such a shift, let alone motivate it, would require some consensus on the question of needs. Thus however problematic a universalistic theory of needs may be, the project of ecological justice requires us to eschew a wholly relativist and conventionalist position,

3

she insists, for it cannot be squared with an unlimited respect for cultural plurality and difference.

In discussing the various problems raised by the attempt to construct such a theory, Soper also takes issue with Benton's view that an appropriate view of human needs can or should be constructed on a naturalistic basis which emphasises human-animal continuity. On the contrary, she believes, the specificity of human flourishing must be understood in its own terms: both the malleability of human nature and its difference from the rest of nature need to be fully acknowledged. This also causes her to be more sceptical than Benton about human obligations to non-humans – these are questionable insofar as they require us to abstract from human interests or conflict with human values of self-realisation. Furthermore, giving due consideration to animals can mean precisely not effacing their differences from us, but fully recognizing and respecting their own, different, ways of life. This is not to say humans should not try to understand and promote the good of other animals, but simply to acknowledge that there are limits to our possible knowledge of what is good for them. Our bonds with non-human nature cannot be indefinitely extended and we cannot relate to it other than anthropocentrically in some sense.

The difficulties Soper identifies with extending present human obligations towards non-humans and future generations of humans are further discussed in the next four chapters: Cooper and Hayward on non-humans; Attfield and Oksanen on future generations.

David Cooper provides further reasons for scepticism about whether the vocabulary of justice is appropriate for addressing issues about the proper treatment of non-human beings. He develops a critique of what he calls the 'mainstream' approach in environmental and interspecies ethics. This approach is characterised by a conviction that the extension of moral concern from humans to non-humans is required for the sake of *consistency*; only if the differences between the human and the non-human were relevant to moral concern would it be consistent and rational to limit it to humans; when they are not, allowing them to influence moral judgement is criticised as 'speciesism'. Cooper credits this approach with an elegant simplicity: it minimises the reliance upon anything but reason, in the strict sense of respect for consistency; and once moral principles regarding human relations are in place, then reason – 'together with a modicum of knowledge about animals (that they suffer, say) and nature (that plants are self-sustaining organisms, say)' – is sufficient to warrant their exercise beyond the human community. However, he proceeds to level two criticisms against this approach.

His first objection develops the point, often made in criticisms of universalistic contractualist theories like Rawls's, that the application of principles of justice involves substantive conceptions of the good and that because different people can have different conceptions of the good, their

4

interpretation and implementation of any principle of justice will also vary. Because of this, the appeal to moral consistency in the extension of human norms to the treatment of non-humans cannot achieve what the mainstream approach desires for it: its proponents, he argues, do not differ from other people by resolutely and consistently applying principles everyone accepts but rather in how they perceive the limits of extension of those principles.

His second objection concerns what he sees as an obsession with searching for similarities between humans and non-humans. On the one hand, alleged similarities can be too stretched and thin to be morally persuasive. On the other, searching for similarities can blind us to other sources of moral concern for nature which can and ought to be important to us. For instance, a sense of restraint in the face of nature, a fear of taking too lightly or inconsiderably our relations to nature, is ethically appropriate but depends precisely on recognizing our difference from and opposition to the rest of nature, not our similarity to it.

He therefore concludes that whereas it is on the basis of similarity that the discourse of justice is invoked, a different rhetoric is required if the force of these other sources of moral restraint towards nature is to be appreciated.

Tim Hayward shares some of Cooper's concerns about the appropriateness of a discourse of justice to capture what is involved in humans' ethical relation to members of other species. But rather than abandon it, he believes, it should be complemented by elements of a discourse of care. He accepts that consistency alone does not suffice to ground obligations in regard to non-human creatures: there also has to be a willingness and ability to *perceive* morally relevant similarities. The inability to perceive them cannot be criticised as speciesism, he grants, but an unwillingness even to try to do so can still be criticised: overcoming 'human chauvinism', as he calls it, may in large measure require good will, but principles of human benevolence perhaps extend further than Cooper would allow. Hayward does not believe that a failure to see relevant similarities should always be condoned or that human chauvinism be allowed to prevail by default. He claims that the burden of proof should rest with those who would deny similarities when reasonable evidence and argument is presented for them.

Hayward's argument is that an ethic of concern for non-human beings can be articulated in terms of interspecies solidarity: this combines an ethic of justice which, in providing the principle of consistency and the practical possibility of obligations, overcomes speciesism, and an ethic of care which motivates the inculcation of a disposition to recognize morally relevant similarities between humans and non-humans. While it is true that many humans may not spontaneously experience this sense of solidarity with non-human beings, it is also true that they do not necessarily feel solidarity with other humans either; his argument is that in the former case, just as in the latter, what we have is evidence of a need for further ethical development, not an irreducible or irremediable fact about humans. His claim is thus that the requirements of an ethic of interspecies concern are formally the same as those of humanistic ethics and he suggests reasons for believing that the content of interspecies concern can be developed on the basis of a more consistent analysis of the content of inter-human concern.

The question of obligations towards future generations is addressed by Robin Attfield. He focuses in particular on the problem environmental economists and policy-makers face when considering what weight to attach to future costs and benefits in relation to present ones. This is the problem of 'discounting' the future. On the one hand, standard discounting procedures assign such small values to even huge future costs that future interests virtually disappear from calculations involved in the decisions that will produce them: this can be seen as a gross injustice in relation to future generations. On the other hand, if future costs are given equal weight to those of the present they are likely to outweigh present interests, which may be unjust for the present generation, may be counterproductive by hampering developments which future generations might benefit from and, if taken seriously, might tend to paralyse action; moreover, present people are unlikely to be motivated to accept this weighting, which means that advocacy of a zero discount rate would do nothing in practice to rectify the injustice of discounting. The difficulties with this question are further complicated, Attfield notes, by numerous other paradoxes and uncertainties regarding not only the magnitude, significance and very existence of future costs, but also the preferences, size and composition of future generations.

Nevertheless, through a critical analysis of a number of arguments, Attfield develops certain provisional conclusions regarding principles and offers constructive proposals for applying them. One principle he strongly argues for is that simply discounting future interests because they are distant amounts to unjustifiable discrimination against them: 'other things being equal, impartiality between times and between generations is morally mandatory, at least where serious interests are at stake.' Some sorts of discrimination, however, are not arbitrary and therefore not unjust: in particular circumstances, special discounts may be applied to take account of factors such as uncertainty, opportunity costs or predictable productivity and inflation. Having defended this basic position against anticipated objections, he considers its practicability. Discrimination against future people is likely to continue, he believes, until they are represented in decision-making; yet given the political will, they could be represented; in his conclusion he introduces proposals for institutional reforms which would make this possible.

Property, rights and environmental values

The question of the opportunity costs to future generations of not developing resources in the present is tackled by Markku Oksanen in a way that highlights the connection with issues of property rights. He examines the ecological implications of John Locke's theory of property and, in particular, of his two provisos on just acquisition: that which prohibits the acquisition of more than one can use without it spoiling; and that requiring 'enough and as good' to be left for others. Oksanen asks whether the spoilage proviso can be harnessed to prohibit unsustainable use of resources. He notes that some critics with ecological considerations in mind think it would place a stringent restriction on resource use, but he also points out that not developing land and resources is itself allowing them to 'waste'. What the proviso mainly expresses, then, is a requirement of productive use of resources. Regarding the sufficiency proviso, Oksanen notes Nozick's point that if applied stringently, every single act of appropriation decreases the chances of others and so none is ever justifiable. Again, though, this interpretation is *too* stringent, since some development of resources can *increase* others' chances, and if this development requires private ownership, then the latter does not necessarily contravene the proviso. Nevertheless, he argues, the sufficiency limitation tells against the justifiability of an extensive privatisation of common resources: in fact, it requires public control of them in order to protect the share of the propertyless, including the propertyless of future generations.

His chapter thus contributes to a critique of the key claim of free market environmentalism, namely, that if all resources are owned privately, the owners will have a vested interest in protecting them from harm. Free market environmentalism is seen by some as potentially the most effective means of securing environmental protection, particularly in view of the problem dramatised in Hardin's Tragedy of the Commons: when resources are owned communally, everyone has an interest in exploiting them as much as possible for their own benefit and has little regard to the harms and costs of doing so since these are distributed more widely. The problem here is that private interests compete in the exploitation of common resources. This could be avoided in at least two ways: one is by regulating the activities of private enterprise in order to plan protection of common resources; the other is to remove the incentive to exploit, indeed the possibility of exploiting, common resources by making them all private property. Free market environmentalism advocates the latter solution.

John O'Neill provides a critique of the approach to valuing the environment which is adopted by environmental economists and free market environmentalists. Their view is that environmental problems arise because environmental goods are free, which means that nobody has to pay anything

7

for their consumption or destruction. Their solution to this is to put a price on such goods so that their wanton use would be curbed. The pricing of environmental goods can be achieved in various ways, by means of either real or 'shadow' markets, but what O'Neill attacks is the very assumption that attitudes and values appropriate to market behaviour are appropriate to valuing environmental goods at all.

He illustrates how this lack of appropriateness is revealed in the work of environmental economists themselves when they attempt to arrive at market valuations by means of 'willingness-to-pay' surveys. When individuals are asked how much they would be willing to pay in additional taxes to preserve some particular environmental good, they are frequently bewildered by what they are required to do. O'Neill illuminates the problem by comparing it with Herodotus's story of a survey conducted by King Darius in which representatives of two different cultures were asked how much money it would take to get them to perform an act which went against their deeply-held cultural values. What the story reveals, claims O'Neill, is that whatever specific values a culture may subscribe to, all cultures hold some things to be 'beyond price'. Environmental goods, he believes, should be considered, as they indeed appear to be by the respondents in the modern survey, as such things. For valuing environmental goods is not a question of what individuals happen to prefer for their personal benefit, let alone what they would be willing or able to pay for their preferences, but a question of cultural values: settling issues arising from competing cultural values is a matter of properly informed public debate, not of competing individual interests as revealed in markets, whether real or shadow.

Neoclassical economics fails because it assumes that price is simply a neutral measuring rod: it is blind to the social meaning of goods in general and thus to how it is part of the inherent meaning of some goods that they are simply 'not for sale', at any price. An appreciation of such meaning in relation to environmental goods is evident in respondents' reactions to the surveys of environmental economists, but the latter systematically attempt to expunge them. If environmental goods are public goods, then the proper way to evaluate them is in public discussion of the value and relative merits of different public projects. O'Neill concludes that putting market prices on environmental goods is not a solution to environmental problems, but a cause of them: commodification of non-market goods – be they human beings and their body-parts or environmental sites – puts them at the mercy of the highest bidder, where in fact they should be protected against any commercial bids at all.

Russell Keat offers a critical response to John O'Neill. While entirely sympathetic to the latter's general intentions, Keat's concern is that the argument as presented may prove either too much or too little: too much in implying that market mechanisms are inappropriate for any use at all, and not just for dealing with environmental goods; too little in failing to account for why the environment should be accorded any non-derivative value, with its protection contingent on the anthropocentric values people happen to accord it.

One objection, then, is that while O'Neill is officially committed to a 'market boundaries' argument, whose aim is to 'keep markets in their proper place', his position is actually liable to collapse into a total rejection of the market. O'Neill's reasons for rejecting the extension of the market to the environment would, if generalised, imply that the market was a wholly inappropriate institution in any context of application at all. This is especially so, argues Keat, in O'Neill's claims about the 'reason-blindness' of the preferences operative in the market: since all such preferences are reason-blind, any objection to the market based on this 'fact' about it would imply that the market was defective *tout court*. To prevent his position collapsing into a total rejection of markets, one would have to show there was something special about environmental decisions that made reason-blindness inappropriate for them but not necessarily for other cases. For instance, if an acceptable justification of reason-blindness is that it allows individuals to be the best judges of their own interests, this justification might not apply to environmental goods, since protecting these might not be in humans' immediate best interests and human individuals may not be the best judges of what is best for environment. However, Keat claims that O'Neill cannot adopt this approach because of his anthropocentrism – and this is his other main objection.

O'Neill does not appear to entertain the idea of any 'good' which is independent of humans' preferences (albeit the considered preferences revealed through public discussion about ends). O'Neill's arguments rely on what Keat regards as a mistakenly anthropocentric conception of environmental value, one which uncritically endorses whatever value is attributed to it through its contribution to humanly valued social relationships. If O'Neill does not see this as a problem, it is perhaps because he assumes that enlightened human self-concern will coincide with what is good for the environment in that living in a right relationship with the environment is itself a central and constitutive feature of human flourishing. Nevertheless, Keat maintains, to sustain this view means adding something that is missing from O'Neill's account – namely, some environmental equivalent of the concern for other persons that is characteristic of relations between humans. The value attributed to the environment cannot consist wholly, as O'Neill implies, in the part it plays in inter-human relationships.

From questions concerning the basis of environmental values and their relation to other human values, the remaining four chapters turn to questions of implementation. That addressed by Donald McGillivray and John Wightman is whether private rights can play any significant role in environmental protection. Acknowledging reasons for scepticism about the view of free market environmentalists that environmental protection is best achieved by providing private rights and liabilities in unowned resources, they nevertheless argue that problems with traditional regulation leave an important role for private law. This role, however, lies not in implementing a market-based approach to environmental decision-making, but in making possible unofficial challenges to official definitions of the public interest: actions in private law have the potential to open up issues about the desirability of a development and its impact on a locality, thereby increasing accountability and participation in the overall decision-making process.

They note current trends in this direction: while private law has traditionally been restricted to the protection of an individual's direct material interests in property, paradigmatically land, looser conceptions of relevant interests are coming to be used in judgements in environmental cases. Particularly interesting in this regard was the EC directive on waste, which dispensed with the nexus with land, providing for strict liability of polluters and yielding rights of action to those suffering injury. This went beyond traditional private rights in two important ways: in allowing liability to attach to any impairment of environment, including anything not in private ownership; and in providing for common interest groups to bring actions.

The advantages they highlight attach not to the privateness of rights per se, but to the uses towards which, suitably modified, they can be put: it is the *unofficial* nature of private law that provides the space for common interest groups, as well as individuals, to argue for a different interpretation of the public interest from that reached by an official body. Indeed, they claim that private action of this sort can foster a wider and more informed debate about a proposed or existing development than may typically be found within routine planning procedures. This does not obviate the need for regulation, and when private law clashes with regulation they believe a twin-track approach which can tolerate different answers is necessary. This protects the democratic legitimacy of legislators and planning bodies against partial interests while at the same time allowing challenges – which may be no less democratic in spirit and intent – to the meaning of the public interest assumed by the official regulatory authority. Thus their case for using private law rests on its potential to counteract regulatory failure by providing an institutional means of opening regulatory decisions to scrutiny. Individuals and groups could privately initiate legal proceedings in which it would be possible to challenge not only the infringement of private interests, but also the regulator's view of where the public interest lay.

Chris Himsworth is more sceptical about the extent to which legal remedies are capable of dealing with accelerating environmental degradation. In fact, he argues, it is difficult to characterise the state of environmental law at present. While bold principles are declared they are poorly articulated in actual legal regimes. He demonstrates this by examining how the 'polluter pays' principle, enshrined in the Maastricht Treaty, is reflected in the law of civil liability for environmental damage in the UK.

In statute law, he finds, provisions are sporadic and unclear: if the environment is polluted, it is uncertain who is permitted to sue the alleged polluter, or how the liability should be quantified, or what kind of harms count as actionable losses. These problems become more acute when it is the 'unowned' environment that is polluted. Relatedly, he also raises the question of whether the kinds of 'non-use' values now discussed by environmental economists (such as option value, existence value, and bequest value) can or should be incorporated into jurisprudence.

The common law tort of nuisance has historically been viewed as particularly relevant for dealing with environmental harms, but the difficulties in applying it, and the array of opinion about its potential application, are such that Himsworth does not think it promising. He does note, however, that leaving the question of compensation aside, this sort of law can have a deterrent function against anti-environmental behaviour. This, though, is first and foremost a political question about what social norms to enforce and how: it involves political judgements about the spread of risks and the role of state and law in this. On a more enlightened view the law of nuisance – invoked necessarily by individuals and, typically, property-holders – will be welcome only if these individuals are capable of acting as efficient surrogates for a wider interest, that of environmental protection for its, not merely their, sake.

Himsworth suggests that the various problems point to a presumption in favour of channelling awards for environmental damages into a public fund – and he notes that some schemes within Europe give direct recognition to this – but it would inevitably be the state, by legislation, which establishes a scheme, rather than judges by development of common law.

The question remaining is whether an all-embracing doctrine of liability is tenable at all. The provisional conclusion arrived at in this chapter is that it is not. For there is no consensus about a workable definition and no sufficiently focused organising ideal to integrate stricter rules of liability and more generous rules of standing into a comprehensive scheme of liability and compensation. Hence the only way forward he sees is in terms of a modest incrementalism of devising specific rules for specific problems.

The specific problem addressed by Anthony Stenson and Tim Gray concerns intellectual property rights (IPRs) in plant genetic resources (PGRs). This is particularly significant as an instance of how questions about property rights in natural resources no longer relate to tangible objects only. Particularly with the advent of genetic technology, different sorts of issues arise. Genetic information itself has become a major productive force capable of yielding huge economic gains and ownership of it has therefore become a hotly contested matter. Who owns this information, and on what basis, is the central issue in the question of intellectual property rights. Stenson and Gray focus

11

on the relation between this recently acquired knowledge and the more traditional forms of knowledge about plant varieties acquired through centuries of selective cultivation. They examine the claim that Western transnational corporations and governments that profit from manipulating life-forms have engaged in unjust use of Third World biological diversity by appropriating germ plasm developed over centuries by local people. Their argument is that whatever injustice there is in North-South relations, it is not appropriately rectified by granting indigenous communities intellectual property rights in plant genetic resources and their associated knowledge.

Stenson and Gray's aim is to show that if 'Third World campaigners wish to argue for IPRs for cultural communities, they should abandon the idea that these communities are entitled *by right* to intellectual property over genetic resources and instead concentrate on consequentialist arguments'. They criticise two types of argument that could be claimed to generate the relevant entitlement. One uses the labour theory of value; but they doubt whether communities are the sorts of entity that *can* labour, in the relevant sense of performing entitlement-creating acts, or whether the development of landraces would anyway qualify as such an act. The other is based on the idea of proximity, so that local communities (and/or states) would own their genetic resources by virtue of occupying the land on which they grow; added to this is the consideration that the transnational entities that have developed new PGRs may thereby obtain a competitive advantage over and thereby harm the original community of proximity. Yet intellectual property rights do not attach to territories in the way that natural resources do, nor do the harms they lead to necessarily affect (only) those from whom original PGRs were taken.

Stenson and Gray also consider the claim that cultural communities are entitled to IPRs in their traditional knowledge, yet because IPRs are strictly limited in time and revert to common property thereafter, knowledge acquired over centuries can effectively be deemed to have reverted to common property. This is a valuable principle, they believe, reflecting the fact that knowledge is not the sort of thing that can be stolen, because the people who generate it do not lose it if someone else acquires it.

Of course, the argument against the entitlement of indigenous communities to IPRs might also be thought to apply, in modified form, against the entitlement of the biotechnology industry itself, at least if the critique of the market advanced in earlier chapters holds good. Certainly, there is an evident injustice in the situation whereby indigenous peoples are not rewarded for their traditionally improved plant varieties while they have to pay market prices for varieties developed from them by the biotechnology industry. Moreover, as well as such direct social costs, there is also the problem of diminishing biological diversity caused by the intensive monocultural agricultural practices which the biotechnology industry so powerfully encourages: modern agriculture, in its heavy reliance on fossil fuel energy inputs has led to a serious reduction in biological diversity in agricultural systems. These are the interrelated problems addressed by Juan Martinez-Alier.

Martinez-Alier evaluates two contrasting approaches to preserving biological diversity with particular reference to current debates around Mexican agriculture, where the conflict is evident between modern agricultural techniques – of so-called 'high-yield' varieties – which are favoured by the market and more traditional methods which can be preserved only through political action. One approach favours a market-based solution which holds that access to natural genetic natural resources should have an economic price (and that 'Farmers' Rights' should be recognised). At present, the market tends to favour modern techniques, but this is because their real productivity is exaggerated: prices of inputs are badly measured and their harmful consequences bracketed out as externalities. The solution would be to cost in ecologically more realistic prices for fuel inputs, loss of diversity and so on, so that ecological agriculture would be chrematistically profitable. But there are serious difficulties – technical and political – in doing this. Market values are distorted by the current distribution of power and income – peasants and indigenous groups would sell their hypothetical 'Farmers' Rights' cheap – simply because they are poor. If the poor sell cheaply, then there is no reason to trust that prices in an ecologically extended market would be an effective instrument of environmental policy. (And while the poor sell cheaply, future generations and other species cannot even come to the market.)

Thus while some argue that an ecologically extended market can incorporate ecological costs into market prices, others argue that the important question is who can best voice the conflict between ecological and economic-chrematistic reasoning. The question is whether genetic resources in general should be commercialised or continue to be 'world heritage'. Political and social movements favour an alternative approach whereby political measures are taken to protect ecological agriculture. Those in favour of ecological agriculture are against IPRs; they do not all even agree with the payment of 'Farmers' Rights'. It is better, they believe, to consider all genetic resources as 'world heritage' protected as necessary by political and legal means. Thus while it may be impossible to arrive at 'ecologically correct' prices, prices may nevertheless be 'ecologically corrected' by means of fiscal incentives. Applying these, of course, depends on political will underpinned by ecological awareness.

The necessity for more determined political will, harnessed to greater ecological awareness, is indeed evident from the various social and legal perspectives provided in this volume: for while there is clearly some potential for economic and legal measures to contribute to environmental protection, a sufficiently comprehensive approach to it will also require an appreciation at all political levels of how the problems of worldwide environmental degradation are inseparable from issues of justice within generations, between generations and also perhaps between species.

Part One
AUSTIN LECTURE

1 Ecology, community and justice

Ted Benton

Introduction

Some advocates of Green politics make the strong claim that it transcends the old divisions between Left and Right. Other social theorists, such as Anthony Giddens and Ulrich Beck, who have attempted to give a central place to social movements and, especially, environmental issues in their work, have come to similar conclusions – most notably in Giddens's *Beyond Left and Right* (1994). For Giddens, reflexivity, detraditionalisation and globalization are three processes whose operation in contemporary societies displaces class politics in favour of what he calls 'life-politics', associated with progressive social movements. In similar vein, Beck (1992) argues that the 'risk' society which has come to replace classical industrial modernity disrupts connections between structures of social inequality and class identities and leads to a reflexive politics concerned with the distribution of 'bads' rather than goods. The lines of cleavage in the risk society do not correspond to the class divisions of industrial capitalism, in part because the scale and scope of the ecological hazards of contemporary societies are universally threatening. The widespread scepticism about science and technology, combined with the growing role of technical expertise in identifying and informing policy-responses to these hazards, gives knowledge a high priority in the new politics.

Few people who seriously reflect on these arguments will deny that the rise of Green social movements and the issues to which they draw attention have important implications for how we construe the political landscape at the end of the twentieth century. My aim here is to think about just how deep those implications go. To anticipate somewhat, I think that the challenge goes very deep indeed, and does require us radically to rethink the established concepts and values of political discourse. However, this is not the same thing as to say that the old rivalries of Left and Right have been superseded. On the contrary, my argument is that in important respects the issues which are posed by contemporary Green and environmental politics confirm and add both force and urgency to traditional diagnoses from the Left. Vast concentrations of unaccountable economic, military, cultural and political power intensify the divisions of wealth and poverty within and between societies and regions, transform and fragment local cultural forms, redistribute the ecological costs of Northern affluence across the globe and threaten military conflict in the scramble for key material resources. Writers from the Left have taken the lead in developing a critical analysis of all this, but in the field of practice, all the traditional vehicles and strategies for radical change on the Left are in deep trouble. What follows is a small part of a wider attempt to do the rethinking which is required of the Left if it is to recover its proper role in giving voice to and mobilising popular resistance. Of course, I hope it will also be of interest to readers who do not share this project: the diagnosis I offer of the implications of the 'Green challenge' is in many respects one which applies across the political spectrum.

The focus of this paper is on the implications of the new Green agenda for one concept in particular: that of social justice. Historically, of course, there has been no essential connection between a concern for justice and the political Left. Pre-modern concepts of justice, indeed, required allocation of goods on the basis of antecedent entitlements, confirming or restoring the status quo. In modern times, too, the Left has no monopoly on the use of the term justice. However, the centrality of questions of distributive justice has come to define the Left in our century to the extent that any challenge to the concept or value of justice must also call into question the continuation of a recognisably Leftist politics. Similarly, any radical transformation in our ideas about justice must call forth a transformation of the Left: and it is to that process of self-transformation that these arguments are addressed.

Of course, the idea of justice is a contested one and this contest itself is one field of contemporary political dispute. Because my aim is a wide and programmatic one, I shall not engage in detailed analytical treatment of rival philosophical positions on justice. Instead I will draw upon and comment upon several views of justice, both as they have been articulated in political philosophy and as they exist in 'common sense' political thought and institutional politics. One important 'broad brush' contrast I shall draw upon is that between formal theories of justice, generally derived from 'first principles' (often associated with liberalism, but sometimes yielding unmistakably socialist implications) and views of justice which at least claim the status of articulations of the normative commitments of actual moral communities. Where I mention names, Rawls will be the main exemplar of the former. Though both of these methods for arriving at concepts of justice may yield socialist conclusions, socialist views of social justice may be further distinguished by the moral priority they accord to the meeting of need (as against 'wants' or 'preferences' as such), and by their sensitivity to the social relational and other contextual conditions for living well.

So, in what follows, I shall begin with four features of the social and political thought which has arisen in association with the Green movement. For my purposes in this context, it makes sense to include the thinking of the animal rights and welfare movements as integral to the 'Green challenge', though in other respects these movements are quite distinct from each other. I then go on to explore how each of these four features of Green thought poses problems for ways of thinking about justice developed (largely) independently of and prior to the rise of contemporary Green movements. The main drift of the argument in each case will be that Green issues do pose deep and serious questions for established views of justice. However, it will also be noted that considerations of justice also pose a challenge to at least some versions of Green social and political thought. In important respects, however, the critical revision required by this encounter between views of justice and Green thought will be seen to confirm, perhaps even intensify, the divisions of Left and Right.

The Green challenge

There are four aspects of the social thought associated with the Green and animal liberation movements which seem to me to be especially challenging. These include both cognitive claims and innovative value-orientations. They are:

1   Natural limits

Humans are held to be dependent on ecological 'life-support systems' which impose outer limits on the scope and scale of human activity in relation to the rest of nature. This is often supplemented by a second claim: that extrapolation of current growth trends (in population and/or resource use, pollution, etc.) into the future predicts the exceeding of natural limits and consequent catastrophic collapse.

2   Human/animal continuity

Since Darwin, this has been the ruling orthodoxy in the life-sciences. The claim is that our species evolved from some apelike ancestor as a result of the operation of broadly similar mechanisms to those which gave rise to other species. It follows that other animal species are our more-or-less distant kin. This basic evolutionary claim can be complemented by subsequent ecological and ethological research to yield two further, but more scientifically contentious, general claims: a) that (many) other animal species have a range of capacities for social interaction, emotional expression, communication and puzzle-solving, and are vulnerable to associated sources of suffering in ways which differ from but are comparable with those experienced by humans; b) that the first claim under (1) above is but a special case of the ecological 'embedding' of all animal species.

3   Non-anthropocentric values

Greens proclaim an innovation in values which marks a qualitative break from all previous Western thought. This innovation is expressed through a variety of different polarities (ecocentric/technocentric; deep/shallow; biocentric/anthropocentric and others), but the core claim is that the non-human world, or parts of it, have value in themselves, independently of their capacity to serve some human purpose. Green thinkers also sometimes claim that substantive normative principles can be derived from nature, or from ecological characterisations of it. A separate but related move is made by animal liberationists who use human/animal continuity to justify extending the scope of received anthropocentric moral theories (mainly rights theory and utilitarianism) beyond the boundaries of the human species.

4   Ecotopia

Distinctively Green visions of the 'good life' are rarely spelled out in terms of detailed institutional forms, but from such speculations as exist, from attempts at actual communal living inspired by Green ideas and from the various 'platforms' and statements of principles issued by radical ecologists we can reconstruct a cluster of widely shared themes and preferences. The Green society would be one in which:

a   humans would live in ways which minimally disrupted the rest of the natural environment;

b   decision-making would be decentralised to small, self sufficient and self-governing communities;

c   self-government would take the form of active, or participatory democracy;

d   either as a separate principle, or as a consequence of the above, these self-governing communities would be radically egalitarian, or 'non-hierarchical';

e   the purposes of individual and collective life would give priority to aesthetic, spiritual and convivial sources of fulfilment, as against the 'materialist' pursuit of material acquisition and competitive advantage;

f   work to meet basic physical needs would be intrinsically fulfilling and employ tools appropriate to small scale egalitarian communities and to ecologically sustainable production.

Social justice and the Green challenge

In what follows, I shall explore some of the implications of these features of Green thinking for views of justice which have been for the most part developed independently of, and prior to, Green thought in its present form. The aim will be to develop a view of justice adequate to the challenge of informing debate about the institutional structures of a just and sustainable society. Although this goes beyond the limits of this paper, such a concept could play a part in the critical evaluation of Green 'ecotopian' thought. I conclude with some very provisional comments which point in this direction.

1   Natural limits and social justice

Taken together, the claims that there are natural limits and that current growth trends are carrying us beyond them towards catastrophe can be used to justify 'going beyond' or displacing concern with social justice in two distinct ways. One argument is that urgent action to secure human (and natural) survival overrides other values. Implicit in this is a widely shared notion of a hierarchy of needs or values, in which survival is the first priority, followed by emotional security and then 'higher' considerations of aesthetics and 'self-actualisation' (important sources in the contemporary environmental debate are A.H. Maslow (1954) and R. Inglehart (1977) – but these writers do not, of course, advocate a 'survivalist' ethic themselves). On the 'survivalist' view it may be necessary to assign authoritarian powers to governments to secure reductions in population growth, or to restrict the ecologically significant uses of certain private as well as common-property and open-access resources in order to protect the environment.

A powerful objection to this way of thinking is to ask why survival is accorded such a high priority. Clearly, there are widely praised and widely condemned acts of self-sacrifice in which war heroes or terrorists willingly die for values or purposes which they presumably set higher than their own survival. It may be retorted that such cases are in one way or another exceptional and inappropriate to set standards for the rest of humanity, for whom survival is the overriding consideration. However, it still seems relevant to ask what, for the great majority of us not given to ultimate acts of

21

supererogation, is the point of 'bare' survival? Survival, I submit, is of value to us solely in virtue of the purposes and experiences it makes possible. Recent public discussion of euthanasia seems to settle upon this as a consensual 'reflective intuition', among secular opinion, at any rate. So, to override all other values in order to secure survival would be perverse. It would be to destroy whatever made survival a valued aim in the first place.

This takes us to the second 'survivalist' position. This is the 'delayed gratification' argument. It relies on the same hierarchy of values and needs as does the first survivalist argument, but the urgency of the ecological crisis is held to be such that if we fail to address it now, we will lose the opportunity to realise other values. This was the core of Rudolf Bahro's early (1982) call for socialists to give priority to the ecological question: there could be no prospect for socialism unless urgent action were taken to secure survival! Although, in Bahro's case, there was a convenient complementary argument to the effect that the objectives of the Greens could only be met by the overthrow of capitalism (which Bahro at that time took to be equivalent to a transition to socialism), there remains a certain echo of earlier postponements until 'after the revolution'. A more recent example is the Campaign for Political Ecology's Manifesto for a Conserver Society (ECO, 1996):

> The struggle to build an ecologically sustainable order must come first, otherwise all other worthwhile goals are doomed. 'Development', be it social or economic, must be subordinated to the overriding priority of protecting the health and integrity of the Earth's life-giving ecosystems.

Despite its considerable polemical appeal, the objections to this sort of approach are also strong. The key problem is that the forms of power relation which are tolerated to address what is seen as an emergency have a habit of persisting as a long term obstacle to the realisation of the promised values. This is commonly acknowledged as an objection to 'vanguardist' strategies on the Left, but it applies equally well to a certain style of ecological alarmist politics which is prepared to countenance extensions of state power without serious consideration of the consequences for other social and political values.

So, even if we take as true the limits/catastrophe claim, it doesn't give us good reasons for either abandoning or delaying the quest to realise other social and political values. However, there are also good reasons for scepticism about the limits/catastrophe claim itself. I have argued elsewhere (Benton, 1989) that ecological limits are a function of the specific articulation of socioeconomic forms with their ecological conditions and media. If this argument is right, we cannot establish 'limits' in abstraction from social relations and processes. It follows from this that the kinds of 'physicalist' modelling which grounded the 'limits to growth' argument are inappropriate. New technologies and forms of social organisation, some of which may be in principle unpredictable, may enable continued growth in agricultural production, industrialisation, population and so on without global catastrophe. This counter-argument to the strong 'limits' position is the basis for more moderate forms of environmentalism which go by such names as 'sustainable development' and 'environmental modernisation'. Of course, it does not follow that such positions are either coherent or feasible. The argument that 'limits' should be theorised in ways which take account of the relationship of social practices to their material conditions and consequences does not do away with the concept of ecological limits as such – it rather relativises them to specific socioeconomic and technical relations and dynamics. Whether currently prevailing power relations and economic structures are capable of being reformed to render them ecologically 'sustainable', and what would be the consequences of such reforms for other values and purposes, remain very much open to question (see, for example, O'Connor, 1994).

Nevertheless, the notion of sustainable development, in almost all its versions, is sharply opposed to 'survivalism'. Considerations of social justice, especially, are integral to virtually all influential definitions. There are two reasons for this. One is that an empirical case can be made out that distributional inequalities are causally responsible for a great deal of environmental degradation. Reducing inequalities, especially by empowering women and reducing rural poverty in the 'Third World' is often held to be a necessary means of achieving sustainability. The second reason is that most advocates of sustainable development include sociopolitical objectives in their view of 'development'– it is held to mean more than mere continued or accelerated economic growth. So justice is a part of the content of 'sustainable development' as a social, economic and political strategy.

However, giving weight to considerations of justice in the context of sustainability poses a significant challenge to established ways of thinking in several respects. The first is the practical challenge that the global context in which the concept of sustainable development has emerged is one in which the agencies charged with developing strategies for implementing it generally lack either the will or the power and resources to do so. But also at the level of national government policy, established ways of legitimating economic inequalities which are invariably associated with capitalist economies are rendered more problematic by the demands of sustainable development. Economic growth – albeit unevenly distributed in both time and space – seems to be intrinsic to capitalist forms of economic organisation. There is also now a widespread acknowledgement that deregulated capitalist growth tends to widen the gap between rich and poor. There are two basic ways in which toleration, even encouragement, of such continuing inequalities can be legitimated in terms of the Rawlsian 'difference principle' (Rawls, 1971). The first is the 'trickle down' model associated with the New Right: unrestrained growth may widen the gap between rich and poor, but the poor are still made better off in absolute terms by the success of the rich. In fact, there is growing empirical evidence that the effects of economic deregulation bring about a reverse redistribution away from the poorest and in favour of the richest. However, the point here is that sustainable development requires re-regulation of capitalist growth, so that both the rate and the character of economic development are brought back within the sphere of economic planning (even if this is done indirectly through fiscal policies). The presumed link between unrestrained growth and rising absolute standards of the poorest can no longer be deployed even as ideology, if sustainability is taken seriously as an objective of economic policy, since support for unrestrained growth is now not itself a legitimate economic ideology.

The other way in which the difference principle can be used to reconcile capitalist growth with justice is the social democratic project of promoting economic growth so as to enhance the tax base for redistributive public provision. Again, how far the poorest actually gain from these redistributive activities is open to empirical question, but for us, the point is that sustainable development calls into question the centrality of growth as such as the central purpose of economic policy. In general terms, drawing on some version of the difference principle to justify the inequalities that characterise capitalist societies requires economic policies which run up against sustainability. The current institutional order might be able to deliver Rawlsian justice, or sustainability, but not both together. This consideration does not, of course, tell against the difference principle itself. However, what it does suggest is that to combine together the requirements of Rawlsian justice and those of sustainability will call for much more profound social change than might have been expected. If we can countenance only economic growth which does not undermine the ecological conditions for future economic activity, questions of distributive justice arise as questions about the mode of wealth creation itself, rather than as add-on optional extras. Since capitalist growth is notoriously generative of distributional inequalities, the 'capitalist growth with redistribution' package which might otherwise be defended in terms of the Rawlsian 'difference principle' becomes much more suspect. The least well-off in this and future generations might well be benefited by a more egalitarian and less dynamic but sustainable economy than by a high growth, inegalitarian, unsustainable one.

Yet another challenge posed for established thinking about justice by the requirement of sustainability is a consequence of economic (and ecological) globalization. Economic forces redistribute materials, manufactured products, foodstuffs, cultural symbols, money, populations, toxic wastes, weaponry, political power and information across the globe in ways which not only fail to correspond with the boundaries of moral and political communities, but which also actually disrupt, fragment and dislocate them. This presents difficulties for any theory of justice which attempts to ground itself in

24

normative beliefs held by geographically localised moral communities. Do communities with the requisite moral consensus and cultural integrity (any longer) exist? Even if this problem can be overcome, the dislocation between the global processes of distribution of goods and bads and the intra-cultural normative order poses other, more intractable ones. How do we think about questions of justice between communities and across cultures? This is a particularly acute problem when a 'good' (e.g., energy use) which comes up for distribution in one community imposes harms (e.g., acidification, fuel shortages, climate change, etc.) on others. In these cases there are questions about distributive justice within the community, but also questions of distributive justice between communities which cannot be settled by appeal to the local normative order of either. To the extent that such global distributions are subject to any normative order at all, it is to the procedural order of international trading agreements such as GATT and NAFTA, which may be more appropriately understood as forms of institutionalisation of economic power relations. Of course, these arrangements may be morally contested in terms of their injustice, but to do so is to suppose either that some form of global moral community is in process of emergence or that some universalistic moral standards may have authority independently of their adoption by some actual moral community (or both).

This latter outcome favours a view of justice such as the Rawlsian one which allows for the critical appraisal of basic social and economic structures in terms of independently grounded moral principles. However, since the distributions which require regulation are global, the social structures at stake are, likewise, global in reach. In Rawlsian terms, a just global economic order would be one which benefited the least well off in the world more than any alternative, so long as it was consistent with basic liberties and equality of opportunity. Again, Rawlsian justice, in the context of sustainability, has very radical implications, in this case for the global economic order.

The final issue posed for our thinking about justice by 'sustainability' is the most widely discussed – the question of distribution across generations. At first sight a communitarian approach might seem well suited to address this issue, since communities are the paradigm repositories of tradition through which moral standards are passed from generation to generation. Strong and stable communities might be expected to have respect for the past and concern for the future. However, again, the conditions under which 'sustainability' is sought are precisely ones under which the requisite characteristics of communities are undermined. Handing on valued goods to one's descendants is a priority which can be given little sense when one's descendants may live out their lives on a different continent and may themselves have quite radically different values and concerns to one's own. The rapidity and unpredictability of cultural change is a particularly difficult problem for any concept of justice concerned with distributions of goods over time. Not so long ago the arable monocultures not far from where I live were orchid-rich downland. Could the farmers who carried out the change of land-use have predicted that a future generation might come to value the unproductive flower-rich grassland more than food surpluses generated by their destruction? For communitarians, what is to count as 'goods', as well as the appropriate norms for distributing them, are given by local cultural practices. Radical cultural change must undermine the conditions of possibility of rules of justice governing allocations over time-spans within which significant cultural change can occur.

Again, only a universalistic view of justice, in which goods could be identified and 'weighted' independently of local cultural norms, could provide decision procedures for cross-generational justice. In the Rawlsian case, the original position must draw a veil over time as well as place. Whether such a construct will bear the weight placed upon it and whether there are feasible alternatives are questions which will need to be addressed.

2   Animals and social justice

I want to start with some reflection on the ontological claims associated with the animal liberation movement. Broadly speaking, animal welfare approaches are linked with an emphasis on sentience as the key 'qualifying condition' for moral standing, whereas the 'rights tradition' is more stringent. Quite complex psychological capacities and vulnerabilities, such that the individual may be considered 'subject of a life' are required to ground rights. Animal rights advocates extend these characteristics to (some) non-human animals on grounds of common-sense experience and usage, as well as evolutionary theory. They are our kin, so why should we suppose that attributes such as consciousness and subjectivity are unique to us?

It is reasonable to believe that individuals of other animal species are the subjects of more or less complex psychological and emotional lives. But evolutionary kinship implies more than this. Organic embodiment and its associated needs and vulnerabilities, sexuality, mortality, liability to disease, capacities and needs for social relations, and complex ecological dependencies are the shared conditions of both human and non-human animal life. More than this, humans and other animals are interdependent in several respects. Most obviously, the majority of human populations directly depend on a small number of other species as sources of food and often shelter and clothing. Where land is managed for forestry products or arable agriculture, populations of other animal species are treated as competitors and eliminated. Urban development destroys habitat for other species of animals and plants on an ever-growing scale. At the same time, areas are deliberately excluded from development as 'reserves' and some animal species are kept as domesticated companions, guards, guides and servants. Finally, other animals have a considerable symbolic significance in most cultures: as sources of entertainment, as means of thought (including thought about our own nature), as objects of worship and superstition and so on.

This list could be amplified, but should be sufficient to at least ground the plausibility of the following propositions:

a   the ecological, social and cultural relations between humans and other animals are bound up with central features of the lives of both. These relations are neither marginal nor incidental. This is true across cultures (though, of course, the nature of the relations concerned varies enormously);

b   hitherto-dominant Western ontologies which divide material beings into 'things' and 'persons' are challenged by contemporary knowledge about other animals. Neither the anthropomorphic assimilation of animals (except possibly a few primates and marine mammals) to personhood, nor their Cartesian reification can reasonably be sustained;

c   at least some non-human animals have enough in common with humans to render the refusal to extend moral standing beyond the boundaries of the human species unreasonable.

The broad perspective I sketch here (and have elaborated elsewhere – Benton, 1993) I call 'human/animal continuism'. From this it follows that if humans are proper subjects of moral concern in their own right, then so are (many) non-human animals. If we also accept that at least one test of the adequacy of any moral theory is that it provides us with means of distinguishing right and wrong in the conduct of central dimensions of our personal and social lives, it further follows that no moral theory which fails to offer such guidance in the sphere of our relations with other species can be adequate. Animals are not only proper subjects of moral concern, but they are also an important test of the adequacy of any moral theory.

Of course, justice is just one moral concept. It may be that animals are morally considerable, but it does not follow that every moral consideration applies to them. Even the most ardent advocate of animal rights will not include freedom of religious confession among their rights. So what about justice? Is the owner of two dogs morally required to treat them fairly? The question is not obviously absurd, so why not ask it in the broader context of evaluating rival possible institutionalisations of our relations with other animals? Can we meaningfully extend the veil of ignorance in the original position to cover even the species membership of the hypothetical chooser of appropriate principles of justice? We can, certainly, as humans, imagine what it might be like to be a veal calf, or a chimp undergoing vivisection, and, indeed, think that institutions which routinise such treatments must be unjust. However, Rawls's method requires us to imagine principles of justice

27

as objects of possible choice by an occupant of the 'original position'. Could a being of indeterminate species occupy such a position? Since the unencumbered self of the original position is at least required to be motivated to the rational pursuit of a life-plan and, indeed, to be able to choose on a rational basis between possible contexts for the living of such a life, we must conclude that animals stand outside the terms of any just settlement. Animals cannot be victims of injustice.

At first sight it may seem that a communitarian approach might be more inclusive of animals. After all, there have been historical societies in which animals could be charged with criminal offences, were entitled to legal representation and so on. Many of the regulations governing our contemporary treatment of laboratory and farm animals, too, could be represented as attempts to apply minimal standards of justice to them. However, this could be used as an argument against the communitarian position in so far as such regulations have their origins not in appeals to existing values of the respective communities, but in moral campaigns to change them. More fundamentally, however, there is a problem about how 'communities' are defined. For Walzer (1983), norms of just distribution are intrinsic to the kinds of goods under consideration, a thesis which depends on more or less consensual cultural valuations of what is to count as a 'good'. Animals do not participate in the symbolic and normative practices through which these goods are identified and valued as such, so it is hard to see how a culture's norms of justice could apply to them.

Both of these negative outcomes are to some degree unsatisfactory or paradoxical. Animals are present in the interstices of human societies and they participate in a wide range of social practices, generally through asymmetrical power relations, but often as active contributors, nonetheless. Much the same could be said of subordinate castes, classes or genders, to whom considerations of justice certainly are due. Similarly with Rawlsian justice. Animals can suffer a range of harms or benefits as a result of different possible 'basic social structures', no less than can humans. It is true that some basic liberties and socioeconomic goods could not be appreciated by animals, but there seems to be no good reason why considerations of justice should not apply to animals over that range of goods which they can appreciate – such as freedom from torture. On the contrary, 'reflective intuition' suggests that they should.

One possible diagnosis of these difficulties, a diagnosis favoured by the perspective of human/animal continuism, is that both liberal and communitarian traditions for thinking about justice arbitrarily privilege certain attributes, either in defining the conditions of admission into moral communities or in characterising the appropriate conditions for deciding moral principles. These attributes include a rather narrow concept of rational autonomy and capacities for participating in the symbolic and normative practices of a human community. These capacities are (probably) uniquely human and have certainly been widely presupposed in Western cultures as definitive of the human/animal boundary. However, the cumulative case for human/animal continuity establishes a dissociation between those attributes which are required for the exercise of full moral agency and those which make their bearers proper objects of moral concern ('moral patients', as Tom Regan (1988) calls them). Now, if non-human animals are proper objects of moral concern, despite not possessing full moral agency, then the main conditions for the appropriateness to them of considerations of justice seem to be satisfied. The pursuit of human purposes is liable to cause harm to non-human animals, either directly or indirectly (e.g., through habitat destruction), so that the condition of scarcity sets their interests in opposition to one another, at least across a range of their mutual interactions. Manifestly, complete benevolence on the part of humans cannot be relied upon in the resolution of these conflicts of interest and neither can adequate knowledge of the interests of animals be relied upon even where benevolence is forthcoming. Prima facie, these considerations seem to call for a view of justice which recognises that non-human animals can benefit from just, or suffer from unjust, treatment, whilst acknowledging that they do not belong to the class of beings capable of establishing or living by the principles in terms of which that treatment is *judged* just or unjust.

Attempting to press some notion of justice beyond species boundaries seems to be supported by strong moral and factual considerations. However, it has some deeply problematic consequences. The de facto inability of practically all non-human animals to participate as moral agents in human moral communities implies an element of irreducible vicariousness or 'paternalism' in any institutionalisation of dispensing justice to them. The interests of non-human animals would, in any such attempt to institutionalise cross-species justice, have to be represented indirectly, by human advocates or 'proxies'. For this to be possible, in turn, human cultures would need to be enlarged by non-anthropocentric understandings of the conditions for flourishing and the vulnerabilities of non-human animals. If the matter of the cultural relativity of 'goods' is a problem for justice across human communities, how much more serious does it become across the boundaries of species?

Another cluster of problems centres on the association (particularly on the political Left) of the concept of justice with egalitarianism. In the animal rights debate, strongly egalitarian positions at the level of abstract principle tend to be softened in the direction of inegalitarian intuitions by the application of secondary principles in actual cases (see Benton, 1993, pp. 86–7). To accept that there are strong reasons for regarding non-human animals as morally considerable is not necessarily to accept that they are equally morally considerable. In the absence of strong arguments in favour of giving equal consideration to the interests of humans and non-human animals, we may be tempted to follow widely shared moral intuitions which recognise the moral standing of non-human animals, but assign a lower priority to it in cases of conflict with human interests (see the discussion of this in Hayward, 1995, chapter 2). There is at least one strong argument for doing this. It is that to render a position of cross-species moral egalitarianism at all plausible, it is necessary to restrict the scope of moral concern for non-humans to a rather narrow circle of species with the requisite similarity to humans in their psychological complexity and vulnerability to suffering. Allowing for some differentiation in the sources and degrees of appropriate moral concern across the species-barrier allows us to bring into the sphere of moral considerability a far wider range of species and forms of social relationship to them.

In their rather different ways, Peter Singer (1976) and Arne Naess argue for equality of moral concern as between humans and non-humans. If successful, their arguments would tell against my proposal of a view of justice across the species boundary which assigns moral standing to non-human beings, but not necessarily equal moral standing. Peter Singer recognises that there are empirical differences of abilities of various sorts between individuals. He seeks to preserve the value of moral entitlement to equal consideration by dissociating it from the empirical characteristics of individuals. Two people of very different levels of ability, for example, should still be given equal consideration as persons. The point can be extended for differences of kind (as distinct from differences of degree, such as ability). So, for example, we might follow Marx in recognising that an egalitarian distribution would take into account differences of need. People, like myself, who have myopic vision need glasses to move around safely; people with normal sight do not. In this sort of case, being able to judge what counts as 'giving equal consideration' to two beings with different characteristics depends on being able to bring to bear background assumptions of the kinds of things both individuals might need or legitimately want to do, and the abilities/resources they would need to do them. That is to say, some notion of a shared 'form of life' is presupposed in judgements as to what counts as 'giving equal consideration'. To the extent that, within the human species, there are commonalities associated with our universal human nature, then equality of consideration can be given a determinate meaning. However, organic, behavioural and ecological diversity across the species-boundary must render the conditions under which the phrase 'equal consideration' can be given a determinate meaning very problematic. The point here is not so much that it may be right to treat humans and non-humans unequally, but more that there may be no clear way of deciding what is or is not equal treatment (though it may still be clear what counts as exploitative, abusive or cruel treatment).

A second attempt to defend cross-species egalitarianism is offered by the 'platform' of deep ecology (Naess, 1990, pp. 28ff.). Naess calls his principle 'biospherical egalitarianism'. He concedes that some 'killing, exploitation, and suppression' is necessary and it is not entirely clear whether he thinks that this constitutes a limit to the practical applicability of his principle, or whether it is included in its meaning. In some ways the latter offers a more promising reading. It might be taken to recognise as a right on the part of humans what is de facto the case for other species: to meet their needs by preying upon or otherwise using or consuming members of other species of animals and plants. The reciprocal of this right would be the non-right of humans to impose harms on other species in excess of what was necessary to meet their needs (or 'vital needs' as Naess expresses it). However, it is clear from the context of Naess's discussion that the distributive justice involved here operates at the level of populations or species. What humans have no right to do is excessively reduce 'richness' (i.e., population size) or 'diversity' (numbers of species). But in the absence of any independent criteria of what count as 'vital needs' in the human case or of the appropriate human population size, the principle seems to generate no clear decision procedures even at the level of populations. The question of criteria of just treatment as between individuals of different species is simply not posed at all.

So far, it seems, there are strong moral and logical considerations which favour the extension of a notion of justice across the boundaries of species, but at the same time, some strong reasons for thinking that an egalitarian conception of justice across species lapses into incoherence. One possible response to this tension might be to abandon the link between justice and egalitarianism. This might be done across the whole scope of the concept of justice, including its use for adjudicating the rival claims of individuals and groups of humans. This would entail a retreat from the whole range of emancipatory uses of the aspiration to justice in modern times and clearly falls out of consideration in the context of the project outlined at the beginning of this paper. A less drastic option might be to retain an egalitarian conception of justice among humans, but to drop the requirement of equal consideration as between human and non-human individuals and populations. This strategy could be rendered less ad hoc than it might at first seem. We might define an inclusive concept (or 'category') of justice which comes into play whenever there is a situation of 'scarcity', competing interests and lack of mutual benevolence and knowledge between individuals or groups which each have moral standing (irrespective of species). This general category of justice imposes a requirement upon all moral agents in such situations to exercise self-restraint in satisfying their wants in consideration of the needs and preferences of competing moral patients. In some contexts this might amount to leaving for them what is their due, in other contexts the more open-ended concepts of diversity and richness as used in the tradition of deep ecology might be more appropriate. In situations where relations between individual or groups of (generally human) moral agents are concerned, then equality of consideration would be the appropriate principle of justice.

Although I think that such a multilayered view of justice might be rendered logically and morally coherent, there remains a danger that the partial abandonment of the tight linkage between justice and egalitarianism might lead to some backsliding on equality among human groups. Pragmatically, it might be preferable to retain the common-sense restriction of the term 'justice' to human relations with one another and invent or adapt another vocabulary (such as 'respect for' or 'responsibility to' the 'other') to express the broader requirement for self-restraint in our relations with other species. Either way, to acknowledge the moral requirement for such restraint is to accept some constraints on legitimate strategies for bringing about justice among humans. These constraints are additional to those mentioned above as implications of the requirement of 'sustainability' and the above arguments for them are an indication of the limitations of the concept of sustainability as a response to Green moral sensibilities.

3 Non-anthropocentric justice?

Ecocentric writers commonly commit themselves to one or other of two sorts of innovative moral claim. The first is that the non-human world (or some parts of it) has value in itself, independently of human purposes. The second is that authoritative norms for human conduct can be grounded in or derived from nature (or from ecological characterisations of it). Thinkers as diverse as Bookchin (1991), Devall and Sessions (1985) and Leopold (1968) have been widely interpreted as making the second kind of claim, though they differ considerably in both the substantive normative conclusions they draw and the interpretations of 'nature' from which they draw them.

This second type of moral claim is the one which looks most threatening to the concept of justice. This is partly because of the substantive moral positions that are commonly advocated by ecocentrics. These invariably assign moral priority to preserving or restoring system-properties, such as (in the case of Leopold), 'integrity, stability and beauty', or diversity, interdependence, cooperation and symbiosis. Now, it is true that justice may also be a property of a system, but it is morally important as a system-property only because the properties of systems have consequences for the wellbeing of members. The moral focus of the concern for justice, at least as it applies to human relations, is the individual subject, whereas the moral focus of the concern for the stability of the 'biotic community' is the biotic community. The moral (as distinct from ontological) holism of most ecocentric moral thinking sets it in opposition to the moralities of rights, justice and liberty which have been at the centre of Western emancipatory thought in modern times. The interesting exceptions here are the anarchist Green theorists, such as Bookchin, who work with a 'mutualist' interpretation of ecological relationships, emphasising symbiotic and egalitarian features of relations in nature, as against the Darwinian picture of nature 'red in tooth and claw'.

Of course, it is quite easy to show that the characterisations of nature from which these normative conclusions are drawn are highly contentious and selective ways of reading into nature the very values which are subsequently read out again. However, it is still worth pausing to consider how well or badly grounded the rival descriptions are. Such features of ecosystems as diversity, stability, and the interdependence of the populations of the different species which make them up are system-level properties which both form the context of and are reproduced outcomes of micro-level processes. These micro-level processes – comprising the interactions between individuals of the same and other species within the localised populations which make up ecosystems – are often 'hierarchical', predatory, competitive, and the lives of individual organisms caught up in them often 'nasty, brutish and short'. Predator-prey relationships, for example, much studied by ecologists, may be quite stable, conducive to diversity, even beautiful, according to taste, when considered at the level of interacting populations, but hardly attractive models for human conduct viewed at the individual level.

It seems, then, that the moral holism of this strand of ecocentric value-theory is not merely a contingent feature. Theorists such as Bookchin who see themselves as belonging to the broad stream of progressive politics can do this plausibly only on the basis of a selective refusal to include the micro-level interactions underlying system properties in nature in the grounding of their prescriptions for human conduct. For more conservative and reactionary thinkers these micro-level mechanisms may well be endorsed and even advocated as means to desirable ecosystem ends.

These considerations can, of course, be bypassed by a reassertion of ethical anti-naturalism. Whatever agreement we might come to about how ecological relations in nature are properly characterised, it would not follow that such a description was in any way normative for human conduct. The logical force of this argument is complemented by the sheer diversity of often contradictory 'oughts' which ecocentric ethical naturalists derive from their various accounts of what 'is'. However, it is important to keep in mind what ethical anti-naturalism does not demonstrate. First, whilst it may show that ecocentric moral claims cannot be logically grounded in factual statements about nature, it does not show that those moral claims may not be supported on other grounds. Stability, interdependence, cooperation and so on may still be desirable features of social, as well as actual features of ecological, systems, without the latter being a reason for believing the former. Second, and more importantly for my argument here, ethical anti-naturalism does not rule out the possibility that factual claims may be relevant to moral questions in a variety of ways. Indeed, any approach to moral theory which seeks to give some place to ethical reasoning, but which acknowledges the force of ethical anti-naturalism, has as its central task consideration of how factual matters may be relevant to without strictly implying moral conclusions.

A view on this, which I have already drawn upon without explicitly stating it, is that a range of factual claims about features of human (and non-human) existence may serve to impose requirements on moral theories, or tests of their adequacy. If all objects of desire were available at no cost and in unlimited abundance, for example, we would have no need for institutions to regulate their distribution, nor principles to guide them, nor moral theories to adjudicate between rival principles. Scarcity relative to the desires of a plurality of individuals establishes the possibility of conflict and competition. Resolution by way of authoritative norms of distributive justice stands as the principal, if not the only, alternative to the Hobbesian options of violent conflict and coercive imposition from above. One requirement, I am suggesting, for any moral theory is that it pass this test: it is a plausible candidate to provide authoritative normative regulation in those areas of social life which require it. My provisional account of what it is for a feature of social life to require normative regulation is that it would otherwise be characterised by violence and coercion. This will certainly do as a first approximation in the case of theories of justice. My claim is not that certain facts about the human condition imply that any particular distribution is just, nor that any particular theory of justice is true. Rather, it is that these factual claims, if true, serve to define the tasks that any moral theory is required to address.

Where does this leave us on the matter of ecocentric values? If we reject, as I do, the view that moral norms can be derived from factual claims about nature or about ecosystems, we are left with the more open question: does what we know about the ecological predicament of human kind define tasks for moral theory which 'mainstream' moral theories have failed to address because of their (relative) lack of ecological awareness? Is, for example, the unencumbered self in Rawls's 'original position' unencumbered in just those respects which need to be included if we are to have a defensible view of the problems a theory of justice has to solve?

To make a start at answering this question, I shall turn to the first sort of innovative ecocentric moral claim mentioned at the beginning of this section. This is that non-human nature (or some parts of it) is valuable independently of its relation to any human purpose. This is sometimes expressed as the 'intrinsic value' of nature, or natural objects. From the secular point of view which I adopt it is hard to make sense of what it could mean for something to have value independently of the human cultural practices through which value is assigned. However, this is not to say that value is a merely subjective matter. Cultural traditions provide rules and resources for the learning, transmission and creation of individual capacities for valuing.

But this intersubjective 'sociologising' of what is independent of individual subjectivity in the making of value judgements does not go quite far enough in two respects. The first, and most controversial, is that there may be some cultural universals in what we value as members of the species. Less controversial, perhaps, is that any serious analysis of cultural traditions of valuing must recognise that they indispensably involve practical engagement with the objects of valuation. Learning how to appreciate music involves listening to it, being taught what to listen for, comparing different performances, trying out different interpretations on different occasions of listening and so on. None of this would make any sense unless what we valued was already there in the music.

My suggestion is that something of this is also true of our valuation of natural objects. In fact the two types of case (art objects and natural objects) are not so easy to separate as it might seem. Most of what we value as 'landscape', for example, is the outcome (both intended and unintended) of past generations of human interaction with the biological and physical environment. Symmetrically, artistic traditions necessarily depend upon and work with the pre-given properties of their materials.

But ecocentrics tend to reserve their deepest regard for nature as untouched by human activity (see Goodin, 1992). For them the highest priority must go not to the 'cultured' landscape, but to the ancient forest, or the Arctic 'wilderness'. Here, the cultural resources through which we appreciate 'wild' nature have no counterpart in the cultural resources through which the object of our appreciation is produced, as in the case of art objects. Nevertheless, these cultural traditions are like those involved in artistic appreciation in that they, too, necessarily require active attention and practical engagement with the object. Gardeners, hunters, anglers, farmers, landscape painters, ecologists, naturalists, ramblers, astronomers, ornithologists, climbers and others come to their different understandings and valuations of the world they engage with through activity and reflection upon it. The ways in which they value the non-human world are certainly products of culture, but what makes that culture possible is that the world has the properties that these practices both discover and represent in their different ways.

The position I am approaching here is a secular reworking of what seems to me true and important in ecocentric value-theory. In the course of practical engagement with, or of intellectual reflection on the world we inhabit, we encounter its autonomy vis à vis human purposes and its awesome transcendence of our cognitive powers. These recognitions can bring us up against deeper questions about the nature of our own lives in the wider context of our presence in a universe of unimaginable complexity and vastness. The proper attitudes of awe and wonder in the face of all this amount to a valuing of the world for what it is, as distinct from valuing it merely as a means to human ends, or as a bundle of resources for economic exploitation.

Some ecocentrics will object to my secular reworking on the grounds that the activity of metaphysical contemplation of our place in nature with which I replace the notion of intrinsic values in nature is itself a human purpose – albeit a very worthy one. This is still a case of valuing nature for some human purpose and so belongs within the spectrum of anthropocentric positions, they might argue. My response is to distinguish between an environmental ethic which advocates preservation of nature because it is necessary to enable humans to engage in metaphysical contemplation and the activity of metaphysical contemplation itself. The former certainly is advocating the protection of nature for a human purpose. However, that human purpose itself necessarily involves a non-instrumental orientation to the non-human world. It is an example of what Max Weber (1978) would have called 'value-rational', as distinct from 'instrumentally rational' action. This suggests that the logical relations between anthropocentrism and ecocentrism are more complex than they are generally taken to be. A certain kind of anthropocentric case is required to secure the conditions for the flourishing of an ecocentric orientation to the world. Reciprocally, the anthropocentric case depends on recognising and valuing the ecocentric orientation to the world as a human purpose.

So, what is the bearing of all this on the concept of social justice? The above reworking of the anthropocentric/ecocentric opposition suggests that there are two dimensions of the problem. First, if we value nature for its own sake, then this implies that protection of nature may set constraints on the ways in which we seek to achieve justice among humans. This might be conceptualised in terms of doing justice as between humans and nature, or we might think of it in terms of the independent value of nature having implications for the ways in which we try to achieve social justice conceived more narrowly as a relation between humans. Either way, there is a second dimension to the problem if we accept that the opportunity to benefit from a non-instrumental (metaphysical, aesthetic, cognitive, etc.) relation to nature is an important good for humans. This second dimension is the matter of how this opportunity is distributed among humans (in this or future generations). It integrates the requirements for protection of nature and concern for social justice, but in a way which deepens the nature-protection requirement beyond what is usual in the more widely diffused concepts of 'sustainability. So far as the first dimension is concerned, we have to consider further the options of a broader conceptual integration of considerations of justice and the protection of nature, on the one hand, as against continuing to think of nature protection as an independent value which might in principle compete with or override considerations of justice among humans.

Our current situation is shaped by the practical consequences of the second option. Unrestrained growth has been advocated as necessary for the resolution of distributional issues. It has been pursued at the expense of non-human nature. At least in their rhetorical strategies, some Third World leaders at Rio insisted on their right to destroy their own ecological heritage in pursuit of living standards comparable to those enjoyed by Westerners. In many local communities in Western countries, blighted by de-industrialisation and high levels of unemployment and social disintegration, the demand for development overrides the desire for environmental quality Opinion surveys show clear evidence of a shift away from environmental priorities in periods of recession. But there are also situations in which powerful minorities are able to impose (non-utilitarian) environmental preservation at the cost of social justice. Colonial powers in Africa, for example, were able to secure the preservation of huge areas of land for their exclusive use as game reserves, excluding indigenous people both from direct use and from any possibility of development. Comparable ecological 'enclosures' have been established in the name of biodiversity preservation in recent times, with the associated exclusion of indigenous peoples. In Western countries, too, some on the Left have argued that pressure groups for rural preservation seek to impose a privileged value-system and lifestyle at the expense of the rural poor.

These historical examples suggest that elite pursuit of environmental preservation at the expense of social justice is counterproductive. If struggles for social justice are experienced as struggles against the privileges of the defenders of the environment, then the environmental case is de-legitimated, as in much of post-colonial Africa. Similarly, pursuit of social justice without concern for the ecological consequences is also liable to be self-defeating. This is partly because of the ecological requirements for sustainability, discussed above, but, more centrally to this part of my paper, also because it involves a disastrous impoverishment of the content of justice itself.

This brings us to the second option mentioned above: to consider some form of conceptual integration of social justice and environmental value. One way of doing this might be to revise a Rawlsian approach by building into the 'original position' a constraint on the selection of principles of justice such that the outcome should be consistent with environmental protection. Indeed, the case for this is conclusive if Rawlsian justice is to be rendered compatible with sustainability. In this case, cross-generational justice requires preservation of those features of the non-human environment which are either goods in themselves or are necessary to the continued provision of other goods.

But does this take us far enough? The requirement of sustainability is an injunction to preserve whatever is necessary for future generations to enjoy the goods we enjoy. However, it says nothing about what goods they might be. We have already seen that Rawlsian justice faces difficulties in offering principles to deal with justice between cultures which value 'goods' differently, and in dealing with comparable issues arising from cultural change across generations. My reworked version of the ecocentric notion of intrinsic value in nature adds some further difficulties. It is not at all clear that the opportunity for the sort of non-instrumental appreciation of nature I tried to describe figures in Rawls's list of primary goods. The socioeconomic goods of wealth, income, power and authority may certainly be dependent on environmental integrity in ways which the sustainability requirement might secure, but they are quite consistent with the 'bundle of resources' view of nature. The basic liberties which comprise Rawls's other category of primary goods do include liberty of conscience, which seems to come close. However, much depends on whether we think of this as a 'negative' liberty, as the non-right of others to persecute us for our beliefs and so on, or as a positive liberty. If we are to think of it as the latter, then the opportunity for a non-instrumental orientation to nature, as a component of human wellbeing, has some very extensive conditions attached to its realisation. It must include freedom from the constraint of material poverty which drives people to destroy their environments out of desperation and freedom from the cultural poverty which excludes from public discourse visions of wellbeing other than what can be promised (if never delivered – see Leiss, 1978) by the commodity system. It must also include the opportunity to participate in shaping the primary institutional forms through which needs are met and the allocation of resources is decided. In other words, it implies the democratisation of both economic and cultural life.

But already in this line of thinking there has arisen a further difficulty for Rawlsian justice. This difficulty is posed when we ask the question, having accepted that the opportunity for a non-instrumental appreciation of nature is to be included as a 'good', what would count as a just distribution of it? If it is included as a basic liberty, then justice is satisfied only by arrangements which give the most extensive opportunity for it consistent with equal opportunity for all. Part of the problem here is that the opportunity for a non-instrumental appreciation of nature is a collective good. The need for it could not be satisfied by dividing up nature into parcels and allocating them to individuals according to a principle of justice. In this respect it is a radically different kind of good from either income or the right to personal property. It is a collective good not just because, as an object of non-instrumental valuing, the non-human world cannot be disaggregated, but also because the opportunity for such an orientation requires access to cultural resources as well. The integrity of human cultures and their non-human environments and sustaining conditions are, on this view, goods in themselves. Justice with respect to them requires both their continuity (though not necessarily unchanged) through time and equality of participation in and access to them (as distinct from shares in their distribution).

4   Green utopia

This discussion, of course, poses more questions than it answers. There is certainly too little here to ground any firm specification of the institutional framework which might give practical effect to a Green view of justice. However, perhaps enough has been done to demonstrate that the considerations posed for political theory by the emergent Green challenge by no means displace the central significance of questions of justice. If anything, the Green challenge requires us to give still more urgent attention to questions of justice, as well as providing new work for the concept and principles of justice to do. At the same time, I have suggested that the Green challenge renders both communitarian and Rawlsian accounts of justice still more problematic than they would otherwise be. The requirement to conceptualise justice across species boundaries is perhaps the most testing of all. So far as the political Left and the more broadly 'progressive' current in modern politics is concerned, this latter challenge seems to require a partial abandonment of the tight conceptual linkage of justice and equality. The more 'anthropocentric' concern for sustainability also constitutes a challenge to the 'growth plus redistribution' strategy for distributive justice which has become widely accepted on the reformist Left.

The above argumentation has also begun to sketch out some of the ways in which inherited notions of justice might be rethought and revised in the face of the Green challenge. In a provisional way this discussion yields the following list of conditions which any 'Green utopia' might have to satisfy in order to be consistent with a 'strong programme' of Green justice:

1   universal freedom from material poverty;

2   cultural traditions offering resources for respecting and valuing the non-human world and equal access to those resources;

3   acquisition of normative control over those economic and technical forces which distribute human life-chances and mediate human impacts on the non-human world;

4   democratic participation in the shaping of cultural and economic life;

5   open access to valued human environments for non-destructive activity and contemplation;

6   preservation of sufficient environmental space in relation to human populations for open access to be compatible with the purposes of appreciation and contemplation of the non-human world;

7   in association with 3 above, procedures for including consideration of the interests of non-human beings in decisions affecting their wellbeing in accordance with culturally formed notions of cross-species justice, or an ethic of responsibility for the 'other';

8   enough long-run continuity and integrity of human cultures to give coherence to the notion of justice across generations.

It is my contention that such a view of justice has a recognisable continuity with the political thought of the socialist Left, especially its 'utopian' strand. However, the Green challenge and the associated changes in moral sensibilities toward the non-human world do demand a radical reworking of that socialist heritage. It is, of course, a further and much more intractable question how such a vision might take root in the broader political culture, and eventually be institutionalised in a realisable 'ecotopia'.

References

Bahro, R. (1982), *Socialism and Survival*, Heretic: London.

Beck, U. (1992), *The Risk Society: Towards a New Modernity*, Sage: London.

Benton, T. (1989), 'Marxism and Natural Limits: An Ecological Critique and Reconstruction', *New Left Review*, 178, pp. 51–86.

Benton, T. (1993), *Natural Relations: Ecology, Animal Rights and Social Justice*, Verso: London.

Bookchin, M. (1991), *The Ecology of Freedom*, Black Rose: Montreal & New York.

Devall, B. and Sessions, G. (1985), *Deep Ecology: Living as if Nature Mattered*, Bibbs M. Smith: Layton.

ECO (1996), *A Manifesto for a Conserver Society*, Campaign for Political Ecology: Leeds.

Giddens, A. (1994), *Beyond Left and Right*, Polity: Cambridge.

Goodin, R.E. (1992), *Green Political Theory*, Polity: Cambridge.

Hayward, T. (1995), *Ecological Thought: An Introduction*, Polity: Cambridge.

Inglehart, R. (1977), *The Silent Revolution: Changing Values and Political Styles Among Western Publics*, Princeton University Press.

Leiss, W. (1978), *The Limits to Satisfaction: On Needs and Commodities*, Marion Boyars: London.

Leopold, A. (1968), *A Sand County Almanac*, Oxford University Press.

Maslow, A.H. (1954), *Motivation and Personality*, Harper: New York.

Naess, A. (1990), *Ecology, Community and Lifestyle*, Cambridge University Press.

O'Connor, M. (ed.) (1994), *Is Capitalism Sustainable?*, Guilford Press: New York.

Rawls, J. (1971), *A Theory of Justice*, Oxford University Press.

Regan, T. (1988), *The Case for Animal Rights*, Routledge: London.

Singer, P. (1976), *Animal Liberation*, Cape: London.

Walzer, M. (1983), *Spheres of Justice*, Blackwell: Oxford.

Weber, M. (1978), *Economy and Society*, Vol. 1, University of California: Berkeley.

Part Two

2 Human needs and natural relations: the dilemmas of ecology

Kate Soper

To avoid possible misunderstandings, I should make it plain at the outset that the discussion here is not directly concerned with the quality, extent or sources of ecological crisis. I am presupposing the agreement of readers to the general truth that the planet is ecologically threatened; that this is primarily the consequence of specific forms of human appropriation of nature (notably those associated with highly industrial and technologically developed societies); and that corrective measures need to be taken. I am also going to assume agreement to the idea that such measures could take a variety of forms and that there is no guarantee at all that the most democratic or socially just will prevail. Ecology is not in itself a politics; nor does concern for pollution and the scarcity of resources imply a commitment to a just resolution of these problems, although – and this a third point on which I am presuming consensus – it ought to: we ought, that is, to be viewing ecological crisis as a collective problem for the global community at large and be seeking an equitable resolution to it.

To put the point more starkly, I am assuming a repugnance to any form of accommodation to natural limits which takes us further along the road of what might be called 'ecological barbarism', by which I mean a condition of relentless competition between nation states or alliances of such states for uncontaminated space and dwindling resources. I say further along that road, because the barbaric route is one we have been travelling for some time in the sense that the more powerful and affluent nations acquired their relative power and wealth by means of colonial expansion and systematic exploitation of global resources and have sought ever since to sustain their privileged status at the cost of the more impoverished sectors of the global community. But the longer we continue along that path, the more acute the competition for resources will become and the more uncivil the methods to which the societies of affluence are likely to have recourse in defending their advantage.

These, we may surmise, could well over time come to include measures which most members of those societies today would regard as deeply immoral (the quite deliberate and cynical manipulation of poverty, disease and famine; the coercion of Third World economies into an almost exclusive servicing of First World needs for bio-fuels and other energy substitutes; ever more fascistic policies on immigration to check the flood of eco-refugees from the more devastated areas of the globe and so on). The road to ecological barbarism is a nightmarish one to contemplate, not least because of the progressive blunting of moral sensibilities that would be the condition of continuing along it. Moreover, it would almost certainly end in genocidal and possibly terminal forms of global warfare. In this sense, a 'barbaric' solution is no solution at all. Only if it is achieved by more equitable and peaceful means can we reasonably speak of a successful accommodation to existing and future ecological constraints.

To argue this, however, is not to suppose that this solution is itself without its problems and I have in mind here not only the well-nigh intractable problems relating to its practical realisation, but the problems attaching to its very concept: the problems of what it would mean to be in a condition of ecological justice and democracy. In other words, granted the commitment to the achievement of such a condition in principle, what are the ethical considerations it brings into play and how are they to be accommodated? It is essentially this question which is the focus of attention, though I would acknowledge that the level at which it is addressed is fairly preliminary and open-ended. My aim here is to orchestrate a problem rather than provide a solution; to bring into view the ethical dilemmas relevant to a discussion rather than predetermine its course.

Let me begin by noting three areas where these may be said to present themselves in acute form.

In the first place, we must recognise that there is no aspiration to ecological justice which does not commit us to some theory of need. To speak of limits and scarcities in nature is to speak of a constraint or lack of resources in relation to some norm of consumption. Much talk of scarcity is clearly either explicitly or implicitly presuming the indispensability, or at any rate the desirability, of continuing within certain patterns of consumption and measures the exhaustion of resources by reference to it: the scarcity of crude oil, for example, is assessed in relation to given structure of needs (for air-flight, motorised transport, plastics and a wide range of industrially produced commodities). Resources, in short, are deemed scarce or abundant in relation to a certain level of historically developed human needs: they are lacking or available for a given lifestyle and much political talk in the West about ecological shortage takes it for granted that the lifestyle in question is that of relative affluence; it assumes this to be most the relevant gauge for judging present and future demands of nature.

But if we challenge this presumption, as many in the Green movement do: if we argue that it is precisely the pursuit of this norm which has been responsible for generating ecological crisis and that it is this which now has to be abandoned or radically revised, in relation to what conception of human needs are we issuing the challenge? Are we assuming here that there is a 'true' gauge of what is needed, or a truly human norm of consumption, in relation to which global resources could in principle be allocated equitably? Does the idea of a global order that is both ecologically sustainable and socially just commit us to defending a universalist conception of human needs, and, if it does, is that reconcilable with a democratic respect for cultural plurality and the relativist conceptions of needs and welfare which have been so strongly advocated in recent challenges to Enlightenment forms of humanism and essentialism?

Secondly, to this let me add, still on the issue of needs, that there are clearly tensions between the demands we place on nature, or the needs we may be said to have in regard to it – most notably between the aesthetic interest in preserving nature as a site of beauty or intrinsic worth – and its use as an essential resource for the satisfaction of other human needs. Policies aimed at preserving the beauties of nature may help to promote a sustainable use of resources and vice versa, so that what practically serves to enhance the aesthetic attractions of the environment may also advance the conservationist cause, but there is no guarantee of this. Preservationists have already complained about energy conversion programmes, such as the use of wind-power, which involve the siting of machinery in some of nature's most rugged and unspoilt reaches. To observe the hands-off approach recommended by some deep ecologists would inevitably be to restrict even the most benign and resource-conservatory interventions in nature. Moreover, the interests we have in protecting the natural environment as a positional good (as beautiful landscape, as a place of retreat, etc.) clearly raise difficult issues from any democratic point of view concerning the control of individual access to this amenity. These issues about our differential needs for nature constitute a further area of ethical tension for the project of ecological justice.

Thirdly, there is the vexed issue of what one might call the 'moral universe' of an ecological ethics: the question of the constituency with which dealings are to be just. One of the main appeals of eco-politics is to 'our' obligations to future generations and this clearly represents a considerable – and inherently problematic – expansion of the usual domain of ethical concern. Some have questioned whether it even makes sense to speak of ethical commitments to those who do not as yet exist, but accepting that it does, there are still questions as to who can be said to have them (can they justifiably be claimed to be the responsibility of humanity at large?). What, we may further ask, are the grounds of any such obligations and over what period of time can they be said to extend?

The question of the scope of the 'moral universe' is further complicated by the fact that the planet is one we share with numerous other living species, all of whom may be said, in virtue of the fact of their natural generation, to have some claim on its territory and resources. At any rate many in the Green movement would insist that we are responsible in some sense for the needs of non-human creatures, even if opinions differ considerably about the quality and extent of this liability. Some have argued that we must observe the rights of at least the 'higher' and more sentient animals to realise the conditions of a flourishing existence; others have suggested that in the case of the great apes we have duties not only to secure their wellbeing but to include them as equals within a common moral community (Cavalieri and Singer, 1993);[1] others again have defended the idea of the equal value of all life forms and called for a biocentric identification with them (Naess, 1989, p. 31; cf. pp. 34–67). There is, in short, a wide variety of Green recommendations concerning our relations to other species life, all of them critical to a greater or lesser degree of an anthropocentric self-privileging and all of them therefore posing the dilemma of what one might call the relative prioritisation of species needs. Whatever we think of these recommendations, (and some of them are clearly very controversial) the questions they raise must be seen to add to the ethical complexities of the project for a just ecology.

In what follows I shall expand on each of these problems in turn, indicating my dissent from a number of ecological conceptions which I believe to be unhelpful in their approach to them.

Human needs and ecological justice

I submit that any just resolution to eco-crisis would require both transnational forms of cooperative policing on such issues as toxic emissions or resource use and dramatic shifts in the pattern of first world consumption – a rejection, to put it in a nutshell, of consumerist perceptions of the good life in favour of modes of existence that place less stress on nature and are consistent with a more equitable distribution of resources. In the last analysis, however, we can only defend the call for either of these moves if we are also prepared to defend the collective nature of certain needs. We could expect no consent to global policing on such issues as ozone depletion, global warming or the protection of particular resources, regions and species, were there no possible agreement on the fundamental conditions essential to any human survival and flourishing; and it is only through an appeal to such conditions that we could begin to justify any call for those nations whose consumption has hitherto played a minimal role in the creation of common ecological afflictions and scarcities to cooperate in the realisation of global programmes to control pollution and conserve resources.

Similarly, any argument to the effect that specific patterns of consumption and their associated conceptions of the 'good life' need to be radically revised as a condition of long-term ecological sustainability, presupposes some consensus on need: it is calling for constraints to be exercised in relation to a certain standard or norm of consumption that could in principle be enjoyed by all of us both now and in the indefinite future. It is operating, that is, in terms of some theory of what would constitute a 'decent and humane' standard of living and assuming its provision for all to be ecologically viable. I would emphasise the abstract nature of these assumptions and I would be the first to grant that there are very real, possibly intransigent, problems standing in the way of any attempt to theorise this norm: to offer a theory of universal needs.[2]

We should not minimise the conceptual problems of defending a universal theory of human needs, or overlook the counterfactual nature of some of its political implications. The only point I am seeking to establish here is the formal one to the effect that the project of ecological justice requires us to eschew a wholly relativist and conventionalist position on needs. It cannot be squared with the argument of those who would insist that there can be no truly just or democratic politics other than through an unlimited respect for cultural plurality and difference. Postmodernist approaches to welfare have insisted that needs are to be viewed as entirely culturally relative and must be conceptualised as the properties of socially embedded individuals; that universalist or so-called 'thin' theories of need are inherently ethnocentric and paternalist in their epistemological claims to have access to the 'truth' about human needing; that only 'thick' approaches to welfare can give proper respect to the experienced status of needs and to the ways in which individuals, or at any rate particular groups and cultures, must be allowed to be arbiters of these and so on. They have, in short, offered an extended critique of the supposedly anti-democratic tendencies of any essentialist approach to needs. But in doing so, I submit, they are defending a conception of democracy from whose perspective it becomes extremely difficult to legitimate any of the policies I am associating with the realisation of ecological justice. At any rate, this is a perspective which would seem to problematise the possibility of transnational ecological agreements and initiatives (since these can be mounted only on the basis of some consensus about the common plight and needs of humanity); and the relativism of its approach would make it difficult to see on what basis we could challenge even the most ecologically exorbitant and wasteful forms of consumption: why should they, too, not be respected as gratifying the experienced needs of particular culturally embedded individuals? Indeed, if 'democracy' is conceived as demanding indefinite respect for the cultural plurality of needs, forms of welfare and conceptions of the 'good' life, it commits us to the absurdity of defending those patterns of consumption which have been most destructive of cultural diversity and are least compatible with its promotion.

Having said that, however, let me add two caveats or qualifications. In the first place I would emphasise that we are committed here only to a minimal essentialism on needs and should be wary of any theory inviting us to overlook the culturally conditioned and mutable quality of human patterns of consumption. Just as the project of ecological justice requires us to accept that human needs are not indefinitely malleable or entirely relative, so it requires us to reject any form of naturalism which emphasises how similarly (rather than differentially) placed we are to other animals in respect of our 'essential' needs and ecological dependencies and seeks to ground ecological policy in this recognition. Being 'green' in our attitudes to non-human nature is often presented as if it involved overcoming human-animal dualism and eroding the conceptual distinctions between ourselves and other creatures. But this is to invite too aprioristic a conception of our needs. We may be similarly placed to other animals in respect of certain basic needs of survival, but we are very unlike them in respect of our capacity consciously to choose our forms of consumption in the light of ecological constraints – and this malleability or underdetermination in respect of our pleasure and modes of flourishing needs to be emphasised as a potential asset of ecological adjustment. For us, unlike other creatures, living in harmony with nature involves rethinking our flourishing in the light of current and likely future resources, the value we place on equality within the human community in the present and the obligations we feel to future generations. Meeting ecological scarcities may quite possibly require us to sacrifice or severely restrict some sources of gratification and self-realisation (very swift and flexible means of transport, for example) which it seems mistaken simply to dismiss as 'false' or 'unnatural' needs. It will certainly require us to be imaginative and undogmatic in our attitudes to what we can enjoy: to open ourselves to the possibilities of an alternative hedonism and to modes of living and self-fulfilment very different from those associated with current assumptions about flourishing. A naturalist approach to our position in the ecosystem which implies that human beings are possessed by nature of a set of needs whose satisfaction is a condition of their flourishing and of which we could in principle gain an objective knowledge, does not necessarily encourage this form of imaginative shift. In other words, to assume too much fixity in what is required as a condition of flourishing would in the end be to undermine the demand that consumption should be accommodated to the limits imposed by nature. At any rate, it would appear difficult on that basis to develop a compelling hedonist case for doing so.

This relates to my second qualification, which concerns the extent to which any restructuring of First World consumption could in principle be democratically achieved: could it amount, that is, to something more than a set of market driven pressures which simply restricted participation in the affluent life while being achieved in ways which avoided any direct political dictatorship of needs. For if we argue that this could not be realised other than through a collectively mandated forms of self-policing, then we are presupposing that significant numbers could over time come to embrace a less consumerist conception of the good life and, in the light of it, voluntarily opt for policies constraining very resource-hungry and exploitative modes of consumption. We are, to put it bluntly, here presupposing the emergence of extensive electoral support for some form of eco-socialist programme. But we cannot even speculate on that possibility, unless we adopt a perspective on needs which gives proper due to the distinctively human capacity to rethink conditions of flourishing.

Any such rethinking on the scale required seems unlikely at the present time. But were it to emerge, it would do so, I suggest, in part in reaction to the negative consequences of consumerist consumption itself and be driven by a hedonist interest in the gratifications promised by less materially fixated lifestyles. (We might detect some embryonic signs of this development in the protests against motorway expansion and similar campaigns.) But it would also surely in part be motivated by more altruistic forms of concern for a more equitable allocation of global resources and for the continued viability of planetary life.

Obligations to the future

In other words, I am assuming that the rethinking in question would be motivated in part by a sense of obligation to future generations and this brings me to one or two points concerning the nature and limits of any such obligation. A first point here is that it cannot currently be imputed (as it often is in discussions of 'human' obligations to the future) to humanity at large. For it would seem that the most compelling grounds for supposing that any given generation of occupants of the planet is ecologically answerable as a collective to all the members of the next lies in the assumption that these latter, too, in virtue of being humans possessed of certain needs for survival and self-fulfilment, have a right to avail themselves, as 'we' have done, of the natural resources essential to meeting those needs. But unless the 'we' of this argument does have the universal applicability it claims, then the attribution of a general species accountability cannot be sustained. If there are some of 'us' who *have been deprived* of the resources whose supposed availability grounds this obligation, then the argument for a collective species responsibility to the future ceases to have the validity which is claimed for it. For the 'human species' to be obliged to the future on these lines is for it also to be obliged to the present: to all those of its members who are currently denied the means of survival, let alone the means of self-realisation or 'flourishing' which, it is assumed, on this argument, are included within the 'legacy' that each generation has a duty to bequeath. In short, there can be no justifiable grounds for arguing that there is a commonly shared 'species' responsibility to ensure ecological sustainability which do not also at the present time provide grounds for insisting that this is a responsibility that has to fall essentially on those sectors of the global community which have hitherto been most irresponsible and profligate in their use of global resources. Thus we can argue that although there *is* an obligation to future generations which is grounded in what is common to us as human beings and in the knowledge we have of ecology, it is precisely because there is that there is also an obligation on the more affluent nations to promote the conditions in the present which might allow it to be more universally assumed and efficaciously pursued: which might actualise what for the time being must remain a merely regulative ideal of 'collective' species responsibility.

But a second point is that any such collective liability would be limited in virtue of the inherent limitations in the knowledge that any given generation can be expected to have about the most long-term consequences of its environmental policies. As various commentators have pointed out, since we cannot predict all possible outcomes of our actions and even those adopted for the best motives can have unintended adverse effects, the notion of human obligations to succeeding generations cannot be construed as extending into an unlimited future (Passmore, 1980, pp. 75–87; Cameron, 1989; Barry, 1977).[3] Just as the notion of a distinctively moral responsibility would begin to collapse were we to regard individuals as accountable for the entire concatenation of effects that might be traceable causally to any of their actions, so it does if attributions of a more collective or generational obligation are interpreted in those terms, as in effect they would have to be if viewed as indefinitely extending into the future. But just as we attribute individual responsibility for the more immediately foreseeable consequences of action, so it can be argued that there is a more general and indefinitely relayed generational liability for predictable consequences of the use, or misuse, of the environment and its resources. Thus it might be argued that the human species does have a continuous intra-generational obligation to ensure ecological viability for those who are not yet born, although there will be definite limits on its accountability at any point in time.

Responsibilities to the non-human

It seems even more clear that there can be no indefinitely extended obligations to other life-forms and that those who suppose otherwise are, at best, deeply confused. Indeed, much of the argument of the 'deep' ecologists (blanket denunciations of human speciesism; invitations to view nature as an autonomous locus of intrinsic value which we should always seek to preserve from human defilement) is deeply questionable insofar as it might seem to require us entirely to abstract from human interests and to give priority to 'other' nature regardless of what merit it might have in our eyes and whatever its ravages upon human health and wellbeing. It is true that those defending a 'deep ecology' perspective of this kind usually do so by reference to human interests: they argue, for example, that human self-realisation is best served by biocentric identification with all life forms. But while this might appear to be a more acceptable formulation of the antihuman speciesist position, it is if anything more confused than any which overtly invites us to subordinate human wellbeing to that of other creatures. For it is clear than anyone arguing for preservation on this basis is operating within a value system that makes it extremely problematic to defend the equal rights to survival of all life forms. Anyone inviting us to view all life as having equal intrinsic value, or deeply to identify with the locust or the streptococcus, the AIDS virus or the mosquito, cannot consistently place more weight on human self-realisation than on the gains that will accrue thereby to any other participant in the ecosystem. Either some parts of nature are more valuable (rich, complex, sentient, beautiful ...) and hence to be more energetically preserved, or they are not. But if they are not, then we should take the measure of the value system involved and not present biocentrism as if it were plainly in the interests of the species being called upon to adopt its values. We cannot both emphasise the importance of human self-realisation and adopt a position on the intrinsic value of nature which would, for example, problematise the use of antibiotics in the prevention of childhood illness.

For the most part, it should be said, those who insist on the 'intrinsic' value of nature do not go to the 'democratic' extreme of pressing the equal worth of all natural entities, but remain committed either explicitly or implicitly to the idea of a hierarchical ordering in nature. Those, for example, defending the cause of animal rights, have challenged the Great Chain of Being conception of human superiority over other mortal creatures and have entered into extensive debate about where exactly one draws the line between those non-humans who may be said to have rights (or, towards whom, at any rate, human beings have special duties), but there is a consensus that we are here talking about rights or obligations which apply only to a restricted range of living beings – those who by virtue of their neurophysiology are capable of a significant degree of sentience.[4] Such animals, whether or not we want to refer to them as 'higher' in virtue of their capacities for feeling (and hence to regard the epithets 'cruel' or 'insensitive' as appropriate descriptions of human maltreatment of them) are clearly being thought of as having claims to moral attention which are denied to less developed forms of life. Even, then, where arguments about the preservation of other species have been couched in terms strongly denunciatory of 'anthropocentric' attitudes to nature, they have very frequently recognised, at least implicitly, that there are logical limits and practical difficulties in pressing to the extreme the case against human perspectives and self-privileging.

But even they, it seems to me, are conceptually confused if they suppose that we could accommodate the interests of non-human animals other than in the light of our own. What needs to be recognised here is that we are inevitably compromised in our dealings with non-human nature in the sense that we cannot hope to live in the world without placing distraint upon its resources, without bringing preferences to it that are shaped by our own concerns and conceptions of worth and hence without establishing a certain structure of priorities in regard to its use. Nor could we even begin to reconsider the ways in which we have been too nonchalant and callous in our attitudes to other life forms, except in the light of a certain privileging of our own sense of identity and value.

This precisely does not mean overlooking differences between ourselves and other creatures. It may on the contrary mean becoming more alert to what is problematic in the attempt to do so. Projects such as that to extend the 'community of equals' to include the great apes may be well-intentioned, but the bonds they seek to cement are arguably too little respectful of the quality of ape life and the ways in which it must be allowed to differ from our own. To argue that fully flourishing animals, who in their normal state have no more capacity than the least self-realised human being to appreciate the meaning of a system of rights and obligations, should be accepted within the human moral community is to overlook some fundamental and critical conceptual barriers; and, indeed, to ignore them at the risk of both abusing members of our own species and failing to protect other species from misguided and potentially harmful forms of protection. For damaged human beings should no more be regarded as comparable to flourishing apes, than should flourishing apes be exposed to the possible forms of maltreatment that might be invited by their legitimation as in some sense human 'equals'. To be sensitive in this area is precisely not to seek to overcome these conceptual barriers or to undermine our intuitive respect for them, but rather to be as open as possible to the implications for non-human nature of the human forms of sensibility with which we are bound to approach it. (And I might add that my points here apply not only to the animal liberationists but also to deconstructionist recommendations: to Derrida's 'yes to the stranger' ethic, whose 'messianic opening' would ask us to preclude all definition of the 'other's' identity (Derrida, 1993);[5] or to Donna Haraway's 'cyborg feminist' injunction to transgress the human-animal-machine schema of conceptual distinctions (Haraway, 1991).)

I am far from implying here that there is something inherently misconceived or presumptuous in the attempt to speak 'on behalf' of other animals. The point is not that we should dispense with human interpretations of their needs (an injunction which it would in any case be impossible to fulfil) and advocates of animal liberation are quite right to highlight the ways in which cruelty or indifference to the consequences of our actions towards other creatures is licensed by particular constructions of human 'needs' or 'identity'. The point is rather that, in any understanding we bring to other animals, we need to be aware of the limits of our understanding;[6] our very empathy with them requires us, as it were, to respect their difference from us and the ways this may constrain our capacity to 'communicate' on their behalf. To 'think' from their position is, as Derek Mahon suggests in his poem, *Man and Bird*,[7] to accept a certain inability to do so:

> All fly away at my approach
> As they have done time out of mind,
> And hide in the thicker leaves to watch
> The shadowy ingress of mankind
>
> My whistle-talk fails to disarm
> Presuppositions of ill-will;
> Although they rarely come to harm
> The ancient fear is in them still.
>
> Which irritates my *amour propre*
> As an enlightened alien
> And renders yet more wide the gap
> From their world to the world of men.
>
> So perhaps they have something after all –
> Either we shoot them out of hand
> Or parody them with a bird-call
> Neither of us can understand.

Considerations of this kind do not imply that we should give up on all attempts to think across the 'gap' between the world of animals and the 'world of men'. To conceive of oneself as an 'enlightened alien' doomed to parodic whistling might precisely count as one such attempt. But they do, I think, mean recognising that our bonds with non-human nature cannot be indefinitely extended and that we cannot relate to it other than 'anthropocentrically' in some sense. Overcoming the damaging forms of separation and loss associated with 'instrumental rationality' does not mean pretending to forms of 'closeness' we cannot have. It means, on the contrary, becoming more stricken by the ways in which our dependency upon its resources involves us irremediably in certain forms of detachment from it. To get 'closer' to nature is, in a sense, to experience more anxiety about all those ways in which we cannot finally identify with it nor it with us. Though in that very process, of course, we would also be transforming our sense of human identity itself.

Notes

1 Their 'Declaration on the Great Apes' calls for 'the extension of the community of equals to include all great apes: human beings, chimpanzees, gorillas and orang-utans' and defines the 'community of equals' as 'the moral community within which we accept certain basic moral principles or rights as governing our relations with each other and enforceable by law'.

2 The most sophisticated attempt to do so, (the theory of 'basic' needs offered by Len Doyal and Ian Gough (1991), is certainly open to the charge that it is too complacently 'Western-ethnocentric' in some of its assumptions about human flourishing. I also think there are huge problems about assuming the *in perpetuo* viability of any norm of needs provision, let alone that of a Swedish level of satisfaction which Doyal and Gough present as optimal. These points are elaborated in my review of their book (Soper, 1993). See also, Doyal's response (Doyal, 1993) and cf. Nussbaum (1992); Drover and Kerans (1993).

3 Cameron and Barry, however, both contest Passmore's conclusion that the basis for our obligations to posterity resides in the love experienced for immediate descendants and cannot be expected to extend beyond that foundation. Commentators, in fact, diverge considerably in the view they take of the implications of the point about limited knowledge for the degree and grounds of responsibility. See also the discussions of Kavka (1978); Attfield (1991, pp. 88–114); R. and V. Routley (1978); and Partridge, (ed.) (1981).

4 For a superb discussion of the debates in this field and critique of liberal approaches, see Benton (1993); for an interesting appraisal of Benton's argument, see Brecher (1994).

5 Derrida here argues (pp. 32–3): '[T]here would be no event, no history, unless a 'come hither' opened out and addressed itself to someone, to someone else whom I cannot and must not define in advance – not as subject, self-consciousness, nor even as animal, God, person, man or woman, living or dead'; and cf. his argument for 'absolute hospitality' in *Spectres of Marx* (Derrida, 1994).

6 For Benton's discussion of the implications of this point, see op. cit, pp. 162–5 and 212–5.

7 From *Four Walks in the Country Near Saint-Brieuc* (Mahon, 1991, p. 16).

References

Attfield, R. (1991), *The Ethics of Environmental Concern*, University of Georgia Press: London.

Barry, B. (1977), 'Justice Between the Generations' in P.M.S. Hacker and J. Raz (eds), *Law, Morality and Society*, Clarendon Press: Oxford.

Benton, T. (1993), *Natural Relations*, Verso: London.

Brecher, B. (1994), review of Benton in *Radical Philosophy*, No. 67, Summer, pp. 43–5.

Cameron, J.B. (1989), 'Do Future Generations Matter?' in N. Dower (ed.), *Ethics and Environmental Responsibility*, Gower Publishing: Aldershot.

Cavalieri, P. and Singer, P. (eds) (1993), *The Great Ape Project: Equality beyond Humanity*, St Martin's Press: New York.

Derrida, J. (1994), 'The Deconstruction of Actuality: an Interview with Jacques Derrida', trans. J. Rée (1993) from 'Passages', *Radical Philosophy*, No. 68, Autumn, pp. 28–41.

Doyal, L. (1993), 'Thinking about Human Need', *New Left Review*, No. 201, pp. 113–28.

Doyal, L. and Gough, I. (1991), *A Theory of Human Need*, Macmillan: London.

Drover, G. and Kerans, P. (eds) (1993), *New Approaches to Welfare Theory*, Edward Elgar: Aldershot and Vermont.

Haraway, D. (1991), *Simians, Cyborgs and Women: the Revinvention of Nature*, Free Association Books: London.

Kavka, G. (1978), 'The Futurity Problem' in B. Barry and R.I. Sikora (eds), *Obligations to Future Generations*, Temple University Press: Philadelphia.

Mahon, D. (1991), *Selected Poems*, Penguin: Harmondsworth.

Naess, A. (1989), *Ecology, Community and Lifestyle*, trans. D. Rothenberg, Cambridge University Press: Cambridge.

Nussbaum, M. (1992), 'Human Functioning and Social Justice: in Defence of Aristotelian Essentialism', *Political Theory*, No. 20, 2, May, pp. 202–46.

Partridge, E. (1981), *Responsibilities to Future Generations*, Prometheus Books: New York.

Passmore, J. (1980), *Man's Responsibility for Nature*, 2nd edn, Duckworth: London.

Routley, R. and V. (1978), 'Nuclear Energy and Obligations to the Future', *Inquiry*, No 21, pp. 133–79.

Soper, K. (1993), 'From Each According to their Need?', *New Left Review*, No. 197, January–February, pp. 113–28.

3   Justice, consistency and 'non-human' ethics

David E. Cooper

1

Doubtless there are many interesting points at which legal and environmental philosophy intersect. The recent explosion in environmental concern makes pressing such questions as whether the notion of a legal subject is unduly extended if legal protection is afforded to trees, or where, if anywhere, responsibility for 'the commons' lies. In this paper, I address a very broad question that arises: is the vocabulary of justice an appropriate one for addressing issues about the proper treatment of animals, non-sentient life and indeed non-living constituents of the environment, such as mountains? Are the wrongs or evils currently done to battery chickens, whales and, perhaps, forests to be conceived as examples of injustice and to be condemned on that ground? I suggest not. My particular target will be what I have elsewhere dubbed the 'mainstream' approach in 'non-human' ethics (Cooper, 1995): for certainly this approach has come to dominate thinking about the treatment of animals and it has at least loomed large in wider debates about environmental concern. I shall not repeat all my reservations about this approach, focusing instead on the pivotal role that considerations of justice play in it.

Before characterizing the mainstream approach, let me illustrate it. First, James Rachels (1989, p. 123) argues that the way to tell whether there is some right animals possess is to identify one humans definitely have and then 'ask whether there is a relevant difference between humans and animals which would justify us in denying that right to animals. ... If not, then the right ... is [one] possessed by animals as well'. Second, Jeremy Bentham (1960, ch. 17, sect. 1), in his famous 'The question is not, Can they *reason*?, nor, Can they *talk*?, but, Can they *suffer*?' passage, implicitly argues that since it is wrong to cause suffering to people, it must be wrong to cause it to animals, since differences between them and us, like the capacity to speak, cannot be morally relevant. If they were, there could be no objection to harming people, like the hopelessly senile, who have lost such capacities. Finally, John Kleinig (1991, p. 76) cites, though without himself subscribing to, the familiar argument that the reason we ought to 'value Nature' is because it too 'possesses those features that serve to mark out Man as a locus of value'.

What is common to these positions is the conviction that moral concern for the non-human (animals, plants or whatever) is required for the sake of consistency with moral concern for humans. Only if the differences between the human and the non-human were relevant to moral concern, would it be consistent and rational to limit it to humans. But they are not relevant: therefore it is arbitrary and irrational so to limit it. Hence the coining of terms, by mainstream thinkers, like 'speciesism' and 'sentientism', by way of analogy with 'sexism' and 'racism', to bring out the arbitrary, discriminatory character of 'anthropocentric' ethics. (So characterized, the mainstream camp also includes Ted Benton, for in his contribution to this volume, his central case against current treatment of animals and the environment is based on its arbitrariness, its lack of rational continuity with our treatment of people.)

Notice that the mainstream approach is compatible with a variety of 'meta-ethical' positions – utilitarianism and rights-based ethics, for example. Whatever the real basis of moral concern for human beings – utility, rights, sympathy – the argument goes that it is wrong, because irrational and arbitrary, not to extend that concern to certain non-human beings or objects, since there do not exist those relevant differences which alone could justify restricting concern to humans alone. It is important to note this hospitable character of the mainstream approach, for it indicates that it is not only those, like Rachels or Tom Regan (1988)s, who ascribe moral rights to non-humans, for whom considerations of justice play the pivotal role in non-human ethics. 'Justice', of course, is a Protean term and in one sense, perhaps, the sphere of justice coincides with the sphere of rights and duties. Injustice is violation of rights. But there is a wider conception, familiar to readers of John Rawls, for instance: justice as fairness. According to that conception, we act unjustly either by acting according to principles that could not have been settled upon under conditions of fair debate, or by failing to apply principles consistently, hence by acting arbitrarily and discriminatorily.

Mainstream thinkers do not (usually) challenge the fairness of the principles which we think should govern our treatment of one another: their complaint is that these principles are not applied consistently. If they were consistently applied, this would enjoin treatment of animals and nature similar to that meted out to people – protection and respect, for instance. It may be that, in failing so to treat non-humans, we violate rights, but this is not an essential component of the mainstream approach. Whether they actually use the term 'justice' or not, all mainstream thinkers should surely be happy to. This is because, of all the virtues, justice is that which is most intimately associated with consistency of treatment. People can be pretty haphazard in their charity, but still count as charitable people. But someone who treats others arbitrarily, whatever other virtues – natural, but sporadic compassion, for example – displayed in treating them, cannot be congratulated on acting fairly.

It is not difficult to appreciate the charm which this approach has for many modern moral philosophers. It minimizes the reliance upon anything but reason, in the strict sense of respect for consistency, when deciding how we should act. (Witness, for example, Peter Singer's insistence, in many of his writings, that his case for animal liberation makes no appeal to sentiment.) It may be that reason alone, *pace* Kant (on one interpretation), is incapable of determining moral principles. Still, on the mainstream approach, once these principles are in place, reason – together with a modicum of knowledge about animals (that they suffer, say) and nature (that plants are self-sustaining organisms, say) – is sufficient to warrant their exercise beyond the human community.

Relatedly, the mainstream approach is consonant with a rather cheerful view of rational moral progress in history. It is frequently found paired with the image of an 'expanding moral circle', according to which morality progresses, primarily, through people recognizing that it would irrational to exhibit moral concern for a given class of creatures without extending it to a further class not relevantly different from the first. The story goes that, with the moral circle now embracing human beings at large – and not just one's fellow-tribespeople or compatriots – people are increasingly appreciating the irrationality of failing to extend it to include animals and, with that accomplished, to non-sentient things of nature as well. For reasons given in my 1995 paper, this image is a travesty of actual history: moral change, for better or worse, obeys no such simple pattern. More importantly, the image is a poor one even when considered as a model for the development of rational moral perception. For example, on every version of the 'expanding circle' thesis I have encountered, moral concern for animals comes in between concern for humans and concern for the non-sentient environment. This fails to fit both very recent history and the moral priorities of very many people. A research project on which I have recently collaborated shows that a considerable majority of the respondents to a questionnaire eliciting their moral attitudes towards animals and the environment are far more outraged by the degradation of the environment (and not for pragmatic, 'anthropocentric' reasons alone) than by battery farming and experimentation on animals.[1] The 'animal issue' which most concerns them, moreover, is that of species conservation – not something which the 'expanding circle' image and the utilitarian or rights-based theories which inspire it, are anyway in a good position to accommodate. (Why, on utilitarian grounds, should killing the last survivors of a species be worse than killing members of an unendangered species? Can a species have a right to exist?) It might, I suppose, be suggested that this only shows that our respondents were not after all rational, since their circle of moral concern does not expand in the way predicted by the mainstream image. But that, I take it, would be a patently desperate ploy to adopt in order to save the image.

2

I want to level two criticisms against the mainstream approach. The first challenges its assumption that there can be a priori criteria, in conjunction with relevant empirical information, for determining when moral principles are being consistently or otherwise exercised. The second questions its urge to base the case for moral concern for the non-human exclusively upon its supposed similarities with the human.

A common criticism of Rawls (1971), which I endorse, is that the contractors in his 'original position', behind their 'veil of ignorance', are not the neutral beings – stripped, for instance, of predilections peculiar to particular societies – that he pretends. For Rawls's preferred principles of justice to emerge from that contract, it is required, for example, that the contractors appreciate the importance and desirability of material goods and self-esteem. Well, go tell that to the Spartans or to Buddhists, with their respective rejections of the value of material goods and of concern for the self. That Rawls does, despite his claims, build into the 'original position' certain less than universal predilections seems to me inevitable. For *any* principles to emerge from the contract, the contractors must be motivated by goals and desiderata which one can at least imagine – and usually more than imagine – that some human beings do not share. To put the point differently: there can be no a priori determination of principles of justice which all rational human beings would settle upon. At most, one can estimate what principles would be agreed upon by people sharing these, but not those, goals and values – people, in Rawls's case, more or less like modern Americans.

Less familiar, however, is another difficulty with a priori determination of what is just and unjust treatment. Once principles have been settled upon – signed up to, as it were – they must be applied. Now how they should be applied and whether a given way of applying them is consistent with the principles themselves, are not matters to be settled a priori. There is a less and a more contentious reason for this. The less contentious one is that principles are subject to implicit ceteris paribus conditions which are not exhaustively specifiable. 'Thou shalt not lie!': indeed, thou shouldn't – but not when the lie would save a life, or when the promisee has reneged on their promises, or when ... ? In other words, *judgement* is required to decide whether, in flouting the letter of a principle, the principle itself has been violated – whether, that is, one has acted inconsistently with it.

Such judgements, in turn, are only honoured or respected, typically, in the context of some consensus on who count as competent judges: a consensus which reflects tradition and custom rather than registering criteria which everyone could settle on by armchair reasoning (see Fleischacker, 1995). The importance of actual historical consensus is, however, still greater than this makes it sound – and this leads me to the more contentious objection to supposing that consistency in application of principles can be determined a priori. Even when a principle seemingly has no ceteris paribus conditions built into it, so that judgement does not seem required in applying it, the question of whether it is being consistently applied on given occasions cannot be divorced, in the final analysis, from the question of how people find it natural to apply it.

This, of course, is to exploit Wittgenstein's point when he says that any rule and any explicit interpretation of that rule can be made to square with any number of different applications. Ultimately the only reason we can give for saying that this rather than that is the proper way of applying it is to say 'This is simply what [we] do.' It is our 'common behaviour', manifesting an 'agreement in form of life', which is the 'bedrock' for regarding these but not those applications of a rule or principle as proper (Wittgenstein, 1969, § 217, §241). The point is not, I should stress, that no applications are ever 'really' consistent or inconsistent with a principle. Rather, it is to emphasize that in so judging it we are necessarily referring to how people sharing a given form of life find it natural or compelling to behave or think.

Both the general point Wittgenstein is making and my exploitation of it in the present context are, to be sure, contentious. Securing the former, which I cannot embark on here, would require a defence of some form of 'nominalism', a rejection at any rate of the 'objectivist' view that our concepts mirror a natural division of the world's furniture into different categories which our language then obediently charts. I do not, fortunately, have to embark on this general strategy in order to lend some plausibility to the point in the present context. If the application of our principles could be determined a priori, we should have to conclude that people who apply them differently are either ignorant, irrational or evil. Either they lack the knowledge necessary to recognize that certain cases fall under the scope of the principle; or they possess this knowledge, but fail to realize what it entails; or they posses both the knowledge and logical acumen, but just don't care about the proper application of the principle.

Now it seems to me that, very often, *none* of these explanations is plausible. This is especially obvious where the people differently applying the principle are widely separated in history or culture. Aristotle and many other Greeks stopped short of applying various principles of treatment to slaves and barbarians; people not so long ago regarded it as perverse to suppose that principles of human respect apply to murderers; Kant and many of his contemporaries did think that lying in any circumstances was a genuine violation of the principle 'Thou shalt not lie!'; and there have been many societies where, unlike our own, principles of permissible sexual behaviour are taken to apply to children as well as to adults. In none of these, nor in numerous other cases, is an explanation in terms of ignorance, illogicality or evil an attractive one. Nor are such cases confined to inter-cultural differences. Some people in our society take it as evident that the injunction against murder applies to foetuses; for others, it is evident that it doesn't. Neither party need be ignorant, illogical or evil and as Alasdair MacIntyre (1981) and others have pointed out, the very interminability of debate on this issue suggest it is incapable of resolution by pondering what is or is not entailed by such injunctions.

Although it is little more than a gesture towards an explanation, we surely do better to think in terms of differences in *perception*, not in knowledge, ratiocinative power and moral fibre, as responsible for such differences in applications of principles. People in the eighteenth century didn't have less knowledge, of a relevant kind, about murderers than we possess: rather, for them, to see someone as a murderer was to see them as someone who, quite patently, has thereby forfeited all rights to decent treatment. People in those cultures where sex with or among children has been permissible have not failed to draw the right conclusion from their knowledge of children: rather, they did not share our current perception of children as creatures fundamentally different from adults in the range of pleasures allowed them.

Whatever the best way to develop this explanatory gesture, rejection of the general adequacy of the explanations offered on the assumption that correct application must be a determinate matter has two important consequences. First, it will make no sense to say of a people with a different form of life and 'perception' from our own that they act inconsistently with some principle – to which they, like us, subscribe – in applying it differently from us. Or rather, if we did say this, we should simply be registering that how they find it natural to apply the principle is not what we find it natural to do. Second, where within a form of life there is no agreement on how it is natural to apply a principle, no shared perception or sense of being compelled to apply it in a certain way, the charge made by some people that other people are not consistently applying it idles.

3

This latter, to return to our topic, is what happens when mainstream theorists charge people with inconsistency and arbitrariness in failing to extend to non-human life principles they accept for the treatment of human beings. Clearly enough, there is no agreement on such an extension: indeed, mainstreamers presumably recognize that they are in a minority in supposing that one does moral wrong to fish by eating them or to forests by felling them. (Recent events in the cattle industry should be enough to have disillusioned any starry-eyed mainstreamer prone to imagine a shared concern for animals. I have yet to see a TV programme on the BSE crisis where anyone so much as mentions the interests of the cows themselves.)

The mainstreamers' charge of inconsistency is made to sound more plausible than it is by misconstruing, indeed parodying, the attitude of 'speciesists' and 'sentientists'. The former, for example, are construed as proposing some one or two differences between humans and animals – such as the capacity for speech – as justification for differential moral treatment. Or, perhaps, as simply saying 'Because they are *human!*' by way of justification. Neither captures the 'speciesist' standpoint: their claim, rather, is that there are countless differences between humans and animals and that it is simply perverse to be asked to overlook all those differences and instead to focus on just one or two similarities as justifying the extension to animals of the same moral concern we extend to one another. It is surely absurd to suppose that people who don't care about how we treat foxes or forests *must* be either ignorant or illogical: to suppose, that is, that either they do not recognize certain obvious facts about foxes (they feel pain, have families, etc.) and forests (they are self-sustaining systems, etc.) or that they do recognize these facts but are so poor at reasoning that they cannot draw the right moral conclusions from them. Rather they are people for whom such facts, weighed against any number of other facts distinguishing humans from non-humans, just don't *count*. (Similarly, to revert to an earlier example, most people in our society today would deny that the many similarities between children and adults count, in comparison with several obvious differences, when it comes to questions of sexual treatment.)

There is no point in the mainstream theorists' insisting that the admitted differences between human and non-human life are not 'relevant' ones, for that it is to pretend to have access to criteria of relevance which transcend what people *actually* find relevant and salient. Nor is there any point in spelling out some principle which everyone accepts in, as it were, capital letters and saying 'Look, if you accept this principle – for example, THOU SHALT NOT KILL JUST FOR GAIN! – you *must* reject treating animals or plants like this.' For the reply will be that no such principle contains within itself, in advance so to speak, how it should be applied. It can be, and often is, a cheap debating point to say to the animal or environmental ethicist, 'How come you don't watch every step in case you kill an insect or a weed?' This is cheap if it implies that the animal or environmental ethicist must be a hypocrite unless they display an extreme fastidiousness in applying their principles that the person making the point is unlikely to display in the case of their own principles. But the point might instead be the valid one that even the mainstream theorist must operate with a sense of, or judgement as to, the sane application of their principles. After all, mainstreamers do not extend the principle of respect for life to the HIV virus, nor even – apart perhaps from the Jains among them – to tsetse flies. Mainstreamers, therefore, do not differ from other people by resolutely and consistently applying principles everyone accepts but, to recall that vague but useful term, in a *perception* of the limits of the extension of these principles.

My aim, I should stress, is not to criticize a person who, familiar with a forest or a mountain, sees it as something which he or she finds it natural or compelling to describe as 'having interests', 'deserving respect', or the like. For what it is worth, I have my qualified sympathies for such descriptions and hence for the attempt to encourage others to share such a perception. My objection is to the charge that people who accept the same facts about forests or mountains, or animals even, must be guilty of irrational arbitrariness if they do not share this person's extended moral vision.

4

I have a further objection – the second line of criticism announced earlier – to the way in which mainstreamers myopically try to get us to share that vision. This, as we have seen, is to get us to appreciate *similarities* between the human and non-human and so to conclude that if it is right to treat people in a certain way, it must also be right to treat non-humans in that way. I do not deny, of course, that it is sometimes apposite and of moral significance to point out some overlooked similarity: it *is* important, after decades of domination by one of the maddest theories ever to have held sway in science – behaviourism – to emphasize the rich character of animals' psychological lives, to remind ourselves that there *is* a way we can know what it is like to be a bat.

But an obsession with searching for similarities is doubly dangerous. To begin with, some of the similarities or analogies alleged must strike many people as so strained or so thin that the likely responses are bound to be 'No, there's all the difference in the world!' or 'OK, I can see an *analogy*, but so what?' One reads, for example, that plants are 'in many ways like ourselves', apparently because they (a) respond to 'environmental circumstances' and (b) are 'pursuing the realization of [their] own good' (Taylor, 1986, p. 154). Likely replies will be that the former is too thin a similarity to have any implications, while the latter similarity is bogus – plants *don't* 'pursue their own good' in any serious sense of those words.

As this example illustrates, those who proclaim hitherto ignored similarities between humans and non-human entities often face a dilemma they rarely confront. Either the terms in which these similarities are alleged are being used in highly extended, figurative ways, in which case it is unclear what follows: or they are being used reasonably strictly, in which case questions that are never actually raised *should* be raised. Take, for instance, references to the 'interests' of forests and mountains. It was claimed, a few years back, that it was against the 'interests' of a mountain in California that a winter-sports resort should be developed on it. If this was just a rhetorical way of saying 'Leave the mountain alone!', so be it: but then the reference to its 'interests' played no genuine role in the argument. If it was meant by way of any serious analogy with human interests, then it was not a joke to ask, as some did, 'How do you *know* that skiing down its slopes is against the mountain's interests? Maybe that's what it's always been waiting for or what God designed it for.' Clearly it is not always in the interests of people and animals to be left alone: hence, prejudice apart, there could be no reason to suppose the contrary in the case of non-sentient nature – not if the term 'interests' is to bear any serious weight (see Ferry, 1995).

The second danger in the obsession with sniffing out similarities between us and the rest of nature in order to motivate a 'rational' extension of moral concern to the latter is that this blinds us to other sources of moral concern for nature which can and ought to be important to us – sources which stem from the radical differences between ourselves and everything else. I shall end by simply offering two such possible sources for consideration. The first is eloquently described by Bernard Williams. He refers to a 'sense of restraint in the face of nature, a sense very basic to conservation concerns'. This, he argues, is 'grounded in a form of fear ... a fear of taking too lightly or inconsiderably our relations to nature'. Far from being 'an extension of benevolence or altruism', such moral restraint will be 'based rather on a sense of opposition between ourselves and nature, as an old, unbounded and potentially dangerous enemy, which requires respect' – respect not in the sense of what we feel towards our fellow human beings, but in the sense in which 'we have a healthy respect for mountainous terrain or treacherous seas' (Williams, 1994, p. 51). Put differently, surely one source of moral concern for nature resides, not in the analogies and similarities between it and us, but in its very Otherness – an Otherness which, to invoke the words of another philosopher, Iris Murdoch (1985), is a proper object of that virtue she calls 'humility', central to which is the capacity to appreciate things precisely because of their distance from the human. Stephen Clark (1997, Introduction) makes a similar point when he writes that this 'too is a moral exercise – to see a thing as Other, and unlike ourselves, and not a creature to be tamed or beaten'.

Finally, as I have suggested elsewhere (Cooper, 1993), the best explanation of our rather selective concern at the current extinction of wild animal species may be that many of these animals – tigers, for example – are especially striking exemplars of dimensions of life increasingly and lamentably absent from modern human existence. Rapid technological change and the 'rootlessness' it brings in its wake – the atrophy of traditions and communities, the breakdown of the extended family, job mobility and so on – mean that we moderns are increasingly without environments: milieux, that is, in which we are thoroughly 'at home', in which we know our way about unreflectingly, and in which we are smoothly integrated. Whatever the final balanced judgement on technological change and its effects, the loss of environments in this sense is something that many people cannot but lament. As Rilke put it (1980, pp. 470–1), unlike animals, 'we always have the look of someone going away'. Unlike those animals, at any rate, which live in the wild, in situ. And that, I suggested, is one reason why we must regret the disappearance of such animals, for with them there also vanish modes of living that were once, in crucial respects, our own. If this is right, then once again it is a difference between animals and ourselves – at least, as we have now become – which is the source of a moral restraint which we ought to exercise: a difference we are unwisely tempted to overlook by those who would ground restraint towards nature on its likeness to ourselves. It is precisely on such grounds that mainstreamers invoke the rhetoric of justice: but it is a different rhetoric which is required if the force of these different sources of moral restraint towards nature is to be appreciated.

Note

1   *Durham University Research Project on Attitudes towards Animals and the Environment*, 1996. Researchers, David E. Cooper, Joy A. Palmer and Jennifer Suggate.

References

Bentham, J. (1960), *The Principles of Morals and Legislation*, Blackwell: Oxford.

Clark, S.R.L. (1997), *Animals and Their Moral Standing*, Routledge: London.

Cooper, D.E. (1993), 'Human sentiment and the future of wildlife', *Environmental Values*, No. 2, pp. 335–46.

Cooper, D.E. (1995), 'Other species and moral reason' in D.E. Cooper and J.A. Palmer (eds), *Just Environments*, Routledge: London.

Ferry, J-L. (1995), *The New Ecological Order*, University of Chicago Press: Chicago.

Fleischacker, S. (1995), *The Ethics of Culture*, Cornell University Press: Ithaca.

Kleinig, J. (1991), *Valuing Life*, Princeton University Press: Princeton.

MacIntyre, A. (1981), *After Virtue*, Duckworth: London.

Murdoch, I. (1985), *The Sovereignty of Good*, Routledge: London.

Rachels, J. (1989), 'Why animals have a right to liberty' in T. Regan and P. Singer (eds), *Animal Rights and Human Obligations*, Prentice-Hall: Englewood Cliffs.

Rawls, J. (1971), *A Theory of Justice*, Oxford University Press: Oxford.

Regan, T. (1988), *The Case for Animal Rights*, Routledge: London.

Rilke, R.M. (1980), 'Duineser Elegien 8' in *Rilke Werke*, Vol. 2, Insel: Frankfurt.

Taylor, P.W. (1986), *Respect for Nature*, Princeton University Press: Princeton.

Williams, B. (1994), 'Must a concern for the environment be centred on human beings?' in L. Gruen and D. Jamieson (eds), *Reflecting on Nature*, Oxford University Press: Oxford.

Wittgenstein, L. (1969), *Philosophical Investigations*, Macmillan: London.

4 Interspecies solidarity: care operated upon by justice

Tim Hayward

Introduction

There is a growing amount of legislation concerning humans' treatment of the non-human world, but little consensus about its normative basis. There are two problems in the way of achieving a consensus which I wish initially to focus on. First, legislation for the protection of non-human nature usually serves, directly or indirectly, to protect certain human interests rather than the interests of the non-human entities ostensibly concerned; secondly, human interests themselves are not homogeneous and conflict in serious ways.

The first problem has occasioned considerable debate within environmental ethics and radical ecological politics, where, as an antidote to 'human-centredness' and instrumental estimates of nature's value, theories of nature's intrinsic value are advanced. Yet whatever the merits of such theories in offering reasons why humans *should* give more than instrumental consideration to non-humans, they are likely to remain uninfluential as long as humans perceive such consideration to cut against their own interests. It does not suffice as a response to this problem to say that humans should simply be less self-interested: for one thing, it is not at all clear that humans could or should disregard their own interests; for another, and this brings us to the second problem, human interests are themselves anything but homogeneous (cf. Hayward, 1997). It is manifestly the case that various human interests conflict as much in relation to matters environmental and ecological as in relation to any other matters. If there is to be any solution, therefore, it will not be achieved by disregarding or attempting to downgrade human interests: pragmatically any attempt to do so would most likely be doomed to failure; ethically it would be unacceptable – since human interests remain matters of ethical concern and it is a mistake to move from an ethical orientation of exclusive humanism to one that excludes humans.

It is therefore worth considering whether a solution could be to extend the existing discourse of fundamental human rights, which is already addressed to the second problem, in such a way as to deal with the first as well. The humans rights discourse, which does enjoy considerable consensus in international law, embodies just the sort of non-negotiable values which seems to be required for environmental legislation. Could it be extended to protect beings other than humans? On one view, as advanced by the Brundtland Report (World Commission on Environment and Development, 1987) for instance, the goals of environmentalism can in fact be presented as essentially an extension of the existing human rights discourse: as well as emphasising the importance of rights of equity and of future generations, the Report endorses as the first general principle of environmental law the fundamental right of all human beings 'to an environment adequate for their health and wellbeing' (p. 348). However, in focusing on the adequacy of the environment *for humans* and in making human needs the criterion of what is desirable and sustainable, this approach can be criticised for allowing insufficient consideration of the rest of nature and especially of other living creatures, in their own right: they are only considered as *resources* for human use. Accordingly, there is a contrasting view according to which it is not simply a case of expanding humans' existing rights to include a right of adequate environmental quality, but of expanding the range of *bearers* of rights to include non-humans.

Whereas constitutional rights have hitherto been restricted to humans, claims that certain primates should also enjoy such rights are currently gaining support; and claims for animal rights more generally are also being advanced; even rights for natural entities other than animals have been canvassed. However, the more the range of bearers is expanded, the more problematic the rights discourse becomes both in principle and in practice: on the one hand, the less the new bearers resemble humans, the more tenuous appears the connection with the reasons which give human rights their moral force in the first place; on the other, this expansion of rights dilutes their force as rights, since there are likely to be so many competing claims that not all can adequately be redeemed, something which downgrades the discourse of rights as such. There are in any case independent problems with the discourse even within the sphere of human relations and these are compounded by trying to increase the range of rights bearers. There are thus reasons for scepticism about whether ecological values can or should be advanced in terms of rights at all.

Nevertheless, if there are strong independent reasons for maintaining the discourse of human rights, then there is also reason for seeking to frame normative principles concerning the treatment of non-humans in terms that are compatible with it. To this end, therefore, rather than reject the rights discourse, or simply attempt to extend its scope, an appropriate approach is to take a step back, so to speak, and ask what makes human rights possible and what gives them their practical and moral force.

Whatever else is involved in the normative underpinning of human rights, their effective existence depends on the willingness and ability of peoples and governments to enforce the obligations that correlate to them; such willingness and ability, I believe it reasonable to assume, presupposes relations and norms of solidarity between human beings. Accordingly, the question to be addressed in this chapter is whether such relations and norms do or could exist between humans and (any) non-humans.

In the first section I explain briefly why I have chosen to investigate the normative discourse of interspecies solidarity in preference to common existing alternatives: that is, why the scope of the ethic encompasses living species and their members rather than either being confined to humans only or extended to 'nature as a whole'; and why the appropriate form of ethic is one of solidarity. The two sets of consideration combine to form the following view of the problem which the discourse is intended to address: practices which allow detriments to other species on the basis of arbitrary distinctions can be called speciesist; the refusal to allow that distinctions are ever arbitrary, because humans are so different from any other species, can be called human chauvinism. Accordingly, speciesism, because of its arbitrariness, can be theorised as a form of injustice; human chauvinism can be characterised as a disposition to see value in humans only. The antidote to speciesism, therefore, can be theorised in terms of an ethic of justice and in section 2 I explain the advantages for interspecies concern of doing so. Nevertheless, just as an ethic of justice is inadequate on its own to capture the full set of human ethical concerns, so it is in relation to interspecies concern. There is therefore the need to complement it with something like an ethic of care and in section 3 I explain how this orientation, in its focus on moral disposition, serves as an antidote to human chauvinism. The argument as a whole thus leads to the conclusion that the requirements of an ethic of interspecies concern are formally the same as those of humanistic ethics and it contributes reasons to believe that the content of interspecies concern, too, can be developed on the basis of a more consistent analysis of the content of interhuman concern.

1   Why a discourse of interspecies solidarity?

My reasons for focusing on the idea of interspecies solidarity can be introduced with reference to the respective strengths and weaknesses of what I have described elsewhere (Hayward, 1995) as reformist and radical conceptions of ecological values.

Environmental concern can be – and on reformist accounts typically is – based on a conception of humans' enlightened self-interest and there is quite a lot to be said for doing so: not only is there the pragmatic strength of grounding environmental obligations in humans' own interests, there is also considerable scope for generating such obligations, since the more humans realise how their own good depends on the good of non-humans and natural relations, the more they will be motivated to promote the latter. Radical critics, however, are not content with this sort of approach: human goods might not in fact coincide so happily with non-human goods and they want to establish independent reasons to care for the environment for its own sake. Nevertheless, many of the hypotheses this leads them to make are implausible or even unintelligible. The idea of caring for *the environment* 'for its own sake' is ultimately unintelligible or ethically and ecologically arbitrary since qua environment, its value can only be estimated in relation to the being whose environment it is. The problem is not surmounted by focusing on nature as a whole rather than 'environment', since such a focus would be quite indeterminate as a guide to action. Nor would the answer be to focus on the good of intermediate totalities like ecosystems, since what makes the condition of an ecosystem good is either as open a question as the value of nature as a whole or, if stated in more determinate terms, as latently instrumental as focus on the environment qua environment. What is not unintelligible or arbitrary is caring for other beings who are liable to harms which are akin to harms humans are liable to and which are of moral concern. Intelligibility is achieved through increased knowledge about the good of those beings; non-arbitrariness is achieved through identifying morally relevant similarities.

Concern for other species generated on this basis goes beyond what might be taken to follow from the principle of enlightened self-interest. That is, it yields not merely prudential considerations such as that biodiversity may prove useful to humans, but the moral view that different species, and members thereof, are worthy of moral consideration in their own right. This is to acknowledge that while what is of concern about the 'environment' qua 'the environment of' may legitimately be a prudential concern, much of the 'environment' nevertheless also consists of living beings who themselves have an environment (of which we in turn are a part) and who may therefore be due a different sort of moral consideration. This draws on the basic organising distinction of ecology between organism and environment. Appropriate norms for the treatment of any environment might be expected to make reference to the interests of the organisms whose environment it is. So is there any reason why only the interests of human organisms should count? What would be the relevant difference between humans and other organisms: pure species membership or specific human characteristics? Principled denials of interspecies norms appeal to the latter. However, I believe these cannot justify denying moral consideration to non-human beings because the characteristics cited are normally those bound up with a capacity for moral agency and those are not necessarily the characteristics which it is the purpose of morality to protect. There is thus no reason to assume that only moral agents can be moral patients. Nevertheless, it might be objected that we cannot extend moral concern to non-human beings because we do not have adequate knowledge about what constitutes their good. Yet I would reply that if any environmental ethic at all is to be possible, our ignorance of the good of other organisms cannot be as radical as it would need to be to refute the possibility of any moral concern with them. For if we are to know how to maintain our environment, then we must know something about what is good for its constituent organisms, at least in the minimal sense of their maintenance in healthy life. I would suggest, though, that in the case of many animals, at least, we know more than this: biologists, ethologists and psychologists can provide quite a bit of knowledge about what is good for the animals they have studied. At any rate, I think we know enough about the good of at least those animals who we involve in our social practices or whose habitats our practices affect to appreciate some similarities with human goods (cf. Benton, 1993). Where such similarities can be established, there is a prima facie ground of moral concern. Of course the moral relevance of the similarities might still be denied, but I believe that the failure to perceive it is due less to inability than to unwillingness to do so. This unwillingness manifests the prejudice of what I call human chauvinism, namely, that membership of the human species is both a necessary and sufficient condition of moral considerability. I call this a prejudice because it disregards how our increasing ecological understanding reveals relationships of interdependence, and even some common interests, between humans and non-human beings.

This, then, is to explain why the focus of normative concern is on interspecies relations. As noted in the introduction, the character of the concern itself is described in terms of solidarity. In the rest of this chapter I shall explain why solidarity in general, and interspecies solidarity in particular, involves both care and justice. My claim, simply put, is that overcoming speciesism is a question of justice and overcoming human chauvinism is a question of care. We cannot overcome the one without overcoming the other. Therefore, just as a complete human ethic involves both care and justice as solidarity, so an ethic of interspecies solidarity involves overcoming both human chauvinism and speciesism.

2 Justice vs speciesism

The normative discourse must incorporate an ethic of justice for a number of reasons. Firstly, due to the fact that any measures for the protection of non-human interests will have to be enforceable, coercion will have to be justifiable, which requires the language of *obligation*. The language of good or value alone will not suffice if potential claims of non-humans are to have any effective force at all, moreover, they will need to be set in the balance against claims of humans. For justice will remain necessary in relation to purely human affairs in any case and any moral discourse developed in abstraction from it would be practically irrelevant. The claims of non-humans have to be rendered commensurable with claims of humans. Hence it is not a question of abandoning the justice orientation, or its expression in terms of fundamental human rights, for instance, but of removing from it the possibility of speciesist interpretations.

The problems with the justice orientation as such, when viewed from the standpoint of ecology or an interest in other species, are connected with how it appears to be structurally confined to – and indeed biased in favour of – specifically human interests. But we need to inquire just why this is so, since it is not obvious why humans' dealings with non-humans could not in principle quite straightforwardly be deemed a matter of justice. There does not appear to be any conceptual obstacle in the way of a society's deeming it wrong, and therefore unjust, for instance, to enclose animals in cruel conditions, to utilise their bodies as resources or objects of experimentation, to expropriate their habitats for human use or to kill them for relatively trivial reasons of human convenience. Some societies have and do recognize such norms. One problem, though, is that such conceptions of justice are *substantive* conceptions: they embody a commitment to a particular theory of what is good and of value; as such, they are ultimately grounded in some religious, metaphysical or ontological account of the world. Such accounts, however, in being based on contestable but unfalsifiable premises, are seen as dogmatic and unpersuasive by modern theorists of justice. Instead, they develop *procedural* accounts of justice.[1] Indeed, questions of justice or right have come to be seen as quite separate from questions of the good or virtue or value. On such conceptions, the *content* of justice becomes a matter to be decided by following the appropriate procedures; those procedures, however, do not determine what the content should be. This, at least, is the official position of proceduralists.

Nevertheless, as various critics have pointed out, the form of justice is not, and cannot be, quite so indifferent to its content as such theorists would have it. Unless the conception of justice is so empty as to have no possible application in the world at all – for example, being restricted to the formal principle of treating like cases alike but with no indication of what counts as the relevant substance of any likeness – any particular conception will involve some substantive commitment. What contemporary proceduralists seek to do, though, is to minimise the commitment, to work with a 'thin' conception of the good. Such a conception can be justified, on a proceduralist account, without introducing strong metaphysical assumptions, by means of a sort of transcendental deduction of the conditions that make justice as such possible: that is to say, it follows from the very meaning of justice as they understand it that there must be rational, autonomous and moral agents capable of forming certain binding agreements between them, therefore it is reasonable to make the prima facie assumption that preservation of the conditions favouring development and exercise of reason, autonomy and morality is a substantive purpose of justice. Unfortunately, from the perspective of our present concerns, these features of agency appear to apply to humans only.

Accordingly, what is at issue is how the form of justice, on proceduralist accounts, is something that appears to be able to govern relations between humans only. Exactly how this is a problem, though, needs closer enquiry.

One aspect of the problem is that the procedures are such that only humans can participate in them. The procedures are designed to discover what norms should have the binding force of obligation on those who are to be bound by them; non-human beings cannot be bound by any obligation and cannot participate in deciding what obligations there should be; therefore non-human beings can participate neither in the formulation nor the application of just principles. Nevertheless, this does not mean that non-humans cannot be beneficiaries or recipients of justice: suffice it to note that some humans are beneficiaries (patients) but not active participants (agents) and therefore non-humans could be too. This aspect of the problem, therefore, would appear not to be decisive.

Another aspect of the problem, though, is that to be a beneficiary of justice is not necessarily the same as having a claim of justice, at least where this is understood as something that can be demanded and redeemed as a right: for a beneficiary may simply be seen as a recipient of charity or a supererogatory act on the part of agents. Certainly this consideration casts doubt on the appropriateness of (at least some kinds of) rights-talk in relation to non-agents. But justice need not be reduced to rights talk; and indeed the inadequacy of the rights discourse, as noted earlier, is one of the reasons for the present inquiry. Nevertheless, people who see in this problem a major stumbling block tend to overstate it. The problem would only be truly serious if having a claim-right were necessary to receiving justice or if being a beneficiary of justice were insufficient for receiving justice. The latter possibility, however, can be disposed of without further ado, it seems to me, since benefiting and receiving justice are indistinguishable, unless recipience is construed in terms of rights – so it is this, the former possibility, that needs to be focused on.

The idea that one must have rights in order to be guaranteed just treatment is one which has considerable practical significance: using a discourse of *fundamental* rights makes possible the identification of specific cases of injustice that need to be remedied. This might be called the signalling function of rights talk. It must be remembered, though, that the point of signalling rights abuses is to invoke obligations to counter them. Thus while in practice identifying rights abuses is important, it is in principle possible to invoke the obligations directly. Incidentally, this also means that there can be a sort of heuristic shorthand use of rights-talk which fulfils the signalling function

73

without implying any legalistic claims: thus all sorts of 'animal rights' issues can intelligibly be identified without any commitment to strong theories of animals as rights-claimants. So the distinction between claims and benefit is not as significant as some would have it. Still, before moving on, it should be noted that the distinction can also be drawn from the side of obligation, as it is by Kant (1964, p. 108) in terms of the difference between an 'obligation to' and an 'obligation with regard to': agents have obligations to other agents to do what their shared conception of justice commits them to; agents do not have any such obligations towards non-agents. Now some critics think this distinction is much more significant than my argument would allow, but as I have already argued elsewhere (Hayward, 1994), the onus is on them to explain what is added to the content of, for example, my obligation 'not to shoot the dog' by calling it an obligation 'to the dog not to shoot the dog'. In both cases the content of the obligation is clear and unequivocal – and the same. It does of course matter *why* there is an obligation to do or refrain from any particular action and that has bearing on who an obligation is due to, but that is a separate matter, which I shall come to next. Meanwhile, I hope to have said enough to indicate why I claim that the beneficiaries of obligations need not be agents or subjects and that nothing therefore necessarily precludes non-humans from being beneficiaries of justice.

Nevertheless, even if there is no conceptual obstacle to considering non-agents as potential recipients of justice, there are difficulties in seeing why agents should accord them this status. What the example of the absent obligation 'to the dog not to shoot the dog' points up is how obligations are agreed to by agents among themselves. Justice, as traditionally conceived, particularly on procedural conceptions, is confined to relations between beings with reciprocal obligations. There is an important principle underlying this conception: obligations are justified if and only if agents agree amongst themselves to be bound by them. On a contractualist view of this agreement there is an element of self-interested consequentialist reasoning: agents agree to be bound by principles of justice, even when it is not in their immediate self-interest, because they wish to preserve the institution of justice itself, since without it there would not be the peace and stability in their social relations which preserve the freedom they are assumed to desire. With regard to non-humans, however, there are no social relations in the first place and there is therefore nothing similar to preserve; there are no threats of civil disorder and so no threats to ward off; there is no possibility of non-humans undertaking any obligations on their part and so no quid pro quo for agents attempting to introduce considerations of them. In other words, the idea of obligations to non-humans appears to be thoroughly unmotivated.

Nevertheless, even if one accepts the contractualist interpretation of the basis of social obligations as described here (chosen because it is the least favourable to my case) there remains more to be said about what it is that the contractors agree to. Justice is, in Rawls's telling phrase, 'the first virtue of social institutions': and the language of virtue here is not fortuitous. Justice does not mean merely keeping individuals from each others' throats in the manner of a Leviathan. Justice is a virtue of institutions not only because it is, as for Rawls, fair, but because it conduces to the good life; indeed, the very idea of fairness is only intelligible against a set of background assumptions about what is good and virtuous. Quite generally, any account of justice which goes beyond the empty formal principle of 'treating like cases alike' to specify, however generally, in what respects cases can be compared and in terms of what criteria, must thereby make some reference, explicit or implicit, to notions of the good. Now the problem in relation to our present concerns appears to be that the notion of the good can be a specifically and exclusively human good. Even on the most formalistic accounts of justice, such as Kant's, the good which provides a regulative ideal for the generation of substantive principles is a specifically human one: for although Kant officially makes 'rational nature' the criterion of good, when it comes to cashing this out he is prepared to speak of 'humanity' as that against which any harm must be deemed a wrong. Thus it is that any obligations which are not held towards identifiable humans are held towards humanity in general. On this account, then, all moral motivations are directed towards the good of humans – whether particular identifiable individuals or humanity in general. Thus, for example, if individuals feel some moral motivation in relation to non-human creatures, as in Kant's (1963, pp. 239–40) example of a man's feeling of obligation not to shoot his dog, this can be explained in terms of wanting to preserve a certain quality of 'humanness' in behaviour and institutions rather than as a manifestation of respect for creatures like the dog in their own right. Yet while this sort of example has been used to illustrate the inherent human-centredness of such accounts of obligations, it does also illustrate something else; namely, that the regulative idea of 'humanity' can include values and motivations that regard the treatment of entities other than humans. The question, though, is not simply whether it can, but whether to any significant extent it actually *does*.

I want now to show why Kant's duty not to shoot the dog is not merely an idiosyncratic or contingent exception to a more generally human-centred rule, but the result of a deep feature of any deontological ethic. On a Kantian account – and indeed, perhaps, on the account of any conscientious rational moral agent – no human action is assumed, without inquiry, to be a matter of moral indifference. Such an account does not start from the assumption that all is license, with any action allowed until proscribed by some obligation. On the contrary, for a rational agent, every proposed action should in principle be examined to see whether it does have any moral implications. Actions can thus be divided into different types: those which are morally justified (shown to be required or permissible); actions that are shown to be morally indifferent;

actions that are assumed, without investigation, to be morally indifferent. The latter are what I call wanton actions.[2] The significance of this is that one does not assume that one can do just as one likes until a reason for providing an obligation is produced. The burden of proof is thus quite different; and this can have quite striking consequences in relation to the treatment of non-human beings. The wanton treatment of non-human beings involves a deficit of justification exactly as any wanton action does. Such actions, moreover, are not only not *justified*, they are not even *motivated* in the sense of being rationally compelling for an agent. Certainly, the features of agency which humans claim as the basis of their alleged prerogatives to treat non-humans just as they list is absent when they do so. There is therefore quite literally no justification for wanton treatment of non-humans.

Yet perhaps there can be detrimental treatment of non-human entities which is *not* wanton? This cannot be ruled out in a world where interests and moral principles can conflict. However, the significance of reversing the burden of proof can be brought out by considering how one particular justification for such treatment can be ruled out a priori and this is the axiom that human interests come first in any case where they conflict with non-human interests. Environmental ethicists sometimes go to considerable lengths to show why this is not an acceptable justification; but their efforts are unnecessary on my view, since it is not, in the first place, a *possible* justification. It is not a possible justification because it appeals to no unambiguous or determinate criterion of judgement. For one thing, human interests are many and various: interests of different humans frequently conflict and some interests of some humans can be bound up with the interests of some ecosystems and some creatures, while interests of other humans are bound up with interests of different non-human beings; therefore determining what is the human priority in one case is only achieved by abstracting from the fact that this may at the same time mean prioritising some non-human interests over different humans' interests. For another thing, and perhaps even more telling, is the problem that 'being human' is not even a determinate criterion: as ethical debates about the moral status of foetuses, human body parts, humans in comas and so on demonstrate, there is not only a lack of identity of interest among discrete humans, there is not even a clear continuity of interest within a 'discrete' human being. If being human is not even a determinate criterion of moral evaluation it cannot be a sufficient one.

No action, therefore, can conceivably be justified on the grounds that it serves human interests in general.[3] Once one seeks to establish more precise criteria, one will find that they do not apply to all and only humans. For instance, if one takes 'rationality' as a criterion, one excludes certain categories of human (including young children and people with impaired faculties) – and it may turn out that one has to include some other mammals too; if one takes sentience as a criterion one includes most animal species; if one takes life as a criterion, one includes all vegetation; if one takes self-regulation as a criterion one includes ecosystems and perhaps the entire biosphere. Environmental ethicists debate at length about which of these criteria to adopt. I advance no particular view on this question, for I have a rather different point to make. Whatever criterion one adopts in any case, one must be consistent in its application – on pain of being a wanton: thus if one believes it is wrong to tie a human up without food or water, because of the pain and distress it causes, then it is wrong to inflict that treatment on non-human creatures who are also capable of experiencing pain and distress in those conditions. This requirement of consistency is deducible from the prohibition on wantonness which is accepted as a necessary condition of moral agency as such.

Once one accepts the consistency requirement, one is committed to avoiding injustices – arbitrary discrimination – in general and thus also the particular form of injustice which can be called speciesism. Speciesism, a term coined on analogy with sexism and racism, means arbitrary discrimination on the basis of species alone when species is not a morally relevant criterion. Humans can appropriately be accused of speciesism when they give preference to interests of members of their own species over the interests of members of other species for morally arbitrary reasons. So, for instance, if it is wrong in the human case to inflict avoidable physical suffering because humans are sentient beings, then it would be morally arbitrary to allow the inflicting of suffering on other sentient beings. That is why cruel and degrading treatment of animals can be condemned as speciesist. More generally, the ignoring of the interests or the good of any being of whom its own good can be predicated can be criticised as speciesist. Avoidance of speciesism, therefore, is revealed as part of one's existing commitment to justice once that commitment is explicated in the manner I have suggested.

Nevertheless, avoidance of speciesism is an insufficient condition for realisation of interspecies solidarity for the same kind of reason that the justice orientation itself is insufficient. The avoidance of speciesism depends on the consistent application of the essential principle of justice which requires that 'like cases be treated alike'. The whole argument advanced so far will hold no sway against someone who simply claims that between humans and non-humans there just are no like cases. If ever and whenever such a claim is true, detrimental treatment of non-humans cannot be unjust or speciesist.

3   Human chauvinism vs care for non-humans

The adjudication that two cases are alike involves not only knowledge – it also involves goodwill and sympathy. There has to be a willingness to *engage* in the comparison, with a genuine openness to the possibility of relevant similarities. Comparing human cases with one another does not usually present insurmountable obstacles: the interchangeability of perspectives has to be assumed for so many reasons that the validity of prima facie likenesses is generally harder to deny than to affirm. The reverse can be the case when it comes to comparisons between humans and non-humans.

Because ascertaining the likeness of cases is a matter of judgement, it will seldom if ever be achieved beyond all possible doubt. Even examples which seem most favourable to doing so can be disputed: for example, while many people have an intuition that the pain and suffering of a cow in a 'factory farm' is relevantly similar to the pain and suffering a human would experience under similar conditions, some nevertheless maintain that they are relevantly different. Such arguments cannot themselves be called speciesist because they involve the claim that there are non-arbitrary reasons for discrimination between the cases: proponents of such arguments could quite consistently avoid speciesist discrimination on the grounds that relevant similarities do not in fact hold. Nevertheless, judgements about the absence of relevant similarities are not entirely beyond possible criticism. It seems to me that the requirement of their being established beyond possible doubt is unnecessarily stringent; I would suggest that the refusal to acknowledge similarities between humans and other creatures can legitimately be criticised if the similarities can simply be established beyond *reasonable* doubt. Thus, if the cases are compared with an open mind, in good faith, it is possible to accept that when the balance of probabilities look favourable to the comparison then it should be assumed to hold; for if avoiding speciesism is reasonable, then taking the course of action least likely to risk it is reasonable. Refusal to take this view is what I refer to by the term 'human chauvinism'.

Human chauvinism is appropriately predicated on attempts to specify relevant differences in ways that invariably favour humans. That is to say, if, when evidence is produced that tends to undermine these claims and assumptions, the response is to seek to refine the definition in such a way as to exclude non-humans once more, then there is a case for thinking this is a human chauvinist response. The case, however, will not always, if ever, be watertight. Human chauvinism is essentially a disposition, and as such requires a kind of hermeneutic to uncover. Thus whereas speciesism can be conceptualised as a clear-cut form of injustice, human chauvinism involves a deeper and murkier set of attitudes. For this reason, ascriptions of human chauvinism are liable to be controversial. Nevertheless, they are appropriate when there is evidence that redefinitions of moral considerability do not simply make more precise the 'rules of the game', but actually involve a progressive shifting of goal-posts in humans' favour. Although it is often likely to be difficult to distinguish between the two cases, evidence of bad conscience and spurious argumentation may sometimes make it less so.

If human chauvinism is essentially a disposition then we cannot expect it to be amenable to correction by justice or coercion. If there is to be an antidote, this too must be dispositional. The antidote, I want to suggest, can be developed along the lines of an ethic of care. Accordingly, I shall first sketch some relevant features of the care perspective in ethics and then explain how it can contribute to an account of a potential normative consensus about the good ordering of humans' interspecies relations.

On my understanding of care, one necessary aspect of it is what might be described as empathy or compassion, or *sympathy* in a broad sense. But while this aspect is necessary, it does not suffice as a basis for interspecies solidarity; one needs to be moved not only in one's emotion or sentiment, one also needs to be moved to action – in other words, practically *motivated*. Thus when one cares, in this full sense, one is not only moved 'inwardly', one is moved to action. Real care, in other words, manifests itself in caring practices: it permeates the core of one's being to the extent of providing a real and effective motivation to act. The motivation will operate when there is the occasion and opportunity to act (i.e., when the harm is being done and when there is something concretely that can be done to combat it); for the rest, care will reside as a *disposition*. An ethic of care, as I understand it, describes morally desirable dispositions an individual should seek to develop.

The idea of an ethic of care has been much discussed in recent years; it has been developed in opposition to ethics of justice and although there is debate as to how sharp and decisive that opposition is, there are nevertheless respects in which it holds good and is relevant to our present concerns. Whereas the ethic of justice is oriented to fairness understood in terms of rights and rules, the ethic of care is oriented to responsibility and relationships; justice aims at universality or impartiality, whereas care aims at preserving the 'web of ongoing relationships'; perhaps most crucially for our present concerns, whereas justice means the non-arbitrary application of *principles*, the ethic of care involves the development of certain dispositions. As I have already argued, proscriptions on speciesism can be formulated as matters of justice and therefore also as laws; but people cannot be forced to overcome human chauvinism, since this requires that they see likenesses with other beings and this they will only do if they are *disposed* to do so.

In this regard, certain questions arise that point to what are already difficulties with the ethic of care in a purely human context and seem to be compounded in the context of interspecies ethics. There are three sorts of difficulty, in particular, that need to be acknowledged. Firstly, if care is extended sufficiently to apply to non-human beings, does it not lose the distinctiveness associated with the emphasis on direct particularity and concrete connectedness, thereby looking increasingly universalistic and even formalistic? Secondly, if care is a disposition, in what sense can it involve obligations? Thirdly, if care is oriented to preserving networks of relationships, how can it apply with regard to beings with whom one acknowledges no relationship? I shall address these three questions in turn.

Firstly, then, is care supposed to be generalizable? Carol Gilligan (1982) speaks of care as a sense of concrete connection and direct response, but she also seems to think of the web of relationships as ultimately encompassing all human beings; other care theorists have affirmed that ultimately each person is connected to each other in virtue of being human. But as critics point out, one thereby loses the contrast between the particularity of relationship and the generality of principle. Once that contrast is lost we seem to be left with an approach that seeks to resolve moral dilemmas through sympathetic identification with all affected parties and this is not much different from what Rawls, for instance, especially on a non-rational choice reading of procedures in his original position (cf. Okin, 1990), advocates in reasoning about justice. Nevertheless, this would only be a problem, I believe, if one's aim was to maintain the hermetic separateness of ethical approaches; if one acknowledges them as two parts of a complete ethic, it is not a problem, since the care element rectifies what has already been described as a defect in the Rawlsian, justice, approach. Indeed, the complementarity is needed on both sides, since this element of the justice orientation is necessary to avoid *dangers* of the care perspective – especially the potential arbitrariness involved in generalising what is particular and context-dependent in ways that overlook precisely the boundedness of its original validity. This brings us into the ambit of the second question.

If care is a disposition, in what sense can it involve obligations? If caring is something one does because one is disposed to, it would seem not only undesirable, but anyway impossible, to force or oblige someone to do it. Indeed, if care were seen as an obligation, the whole distinction between care and justice would be undermined. Still, as feminist discussions of caring have pointed up, it often *is* a matter of obligation: for instance, the patriarchal assumption that mothers and daughters have a disposition to care underlies their effective coercion into caring roles. It is therefore important to distinguish between the caring disposition and caring *duties*. Caring duties can be undertaken even where the disposition to care is absent: for instance, members of caring professions may from time to time exercise their duties with a complete absence of any subjective disposition in favour those they are charged with caring for. So when care theorists argue that we make moral progress by expanding the scope of injunctions to give care, we need to consider what this means. On the one hand, it could be to make a political point – such as, for instance, that men should assume more caring roles: this, though, would seem to be a question of *justice* in the distributions of burdens within families and society at large. On the other hand, if the point is that the amount of caring in the world should simply increase, one must consider why it is that care can be considered a good in itself and what weight it has in relation to other goods. One then notices that what the emphasis on care points to is the value of relationships and of the virtues relevant to their flourishing. These values are established quite generally; therefore the care ethic does not issue in particular obligations to care for specific others in specific ways, but, rather, in a general recommendation to develop a caring disposition. This recommendation would be very much in the spirit of traditional accounts of what constitutes a good life.

It is worth recalling that in its original formulation the care ethic was set out not as a duty but as a description of a moral style actually encountered in empirical research. The care orientation is one which actually exists among human agents; and, according to the findings as reported, it may be more prevalent among females than males. These points suggest an answer to the third question.

The question of most direct concern to *interspecies* care is why people should care about non-human beings with whom they have no networks of interrelationships. Part of the answer is to point out that it is inappropriate, for the reasons just given, to talk about *what* people *should* care about at all. The ethic of care is about identifying and enhancing what people already do care about (although whether they are right to care for whom they do and in the way they do is a question best answered from the justice perspective). To give the answer a constructive turn, therefore, one has to enquire whether people already do care about other beings with whom they acknowledge no relationship. Such an inquiry need not be undertaken in an empiricist spirit. Indeed, having said that the ethic of care is grounded in empirical research, it needs to be added that it has been developed as a *critical* reconstruction of research findings. From the discovery that, empirically, males tend to care less, the hypothesis is advanced that this is not because of biological sex differences, but because the care orientation has been stunted or squashed; the point of care ethics, as I understand it, is to inculcate in all humans a caring disposition. This inculcation will be made possible by cultural changes and these changes given direction by critical knowledge of how ethical dispositions are determined. This sort of critical knowledge, I want to suggest, is what has to be sought in our investigation into the inevitability or otherwise of human chauvinism.

If human chauvinism, as an absence of care about non-humans, cannot be proscribed in the manner of a duty for the same reason that an absence of care in general cannot, nevertheless care for non-humans can be inculcated and there will be grounds for seeking to do so if it can be established that humans already have a disposition for it which is currently disfavoured by particular cultural and ideological factors.

A way of proceeding therefore, analogous to Gilligan's research, is to investigate whether there is already 'another voice' in moral conversations to which dominant accounts of ethics tend to be deaf. One could thus seek to test the hypothesis that we already *do* care in some ways, but do not adequately realise this cognitively (because of repressions, be they cultural, ideological, etc.) and do not adequately act on it (because channels for doing so are blocked off). The hypothesis that our care for other species is actively repressed could be tested by examining, for instance: how children's attitudes towards animals and meat are formed; why food packaging attempts to remove all trace of its animal origins; how attitudes to nature in general as radically other are formed; and so on. There does seem to be a good deal of evidence of our cognitive alienation from the rest of the natural world and, indeed, from our own animal nature (cf. Dickens, 1996). For clues on how to counter this alienation, one might investigate the discourses of the growing number of vegetarians and campaigners for animal welfare with an ear to a 'voice' that others might recognize as their own under more propitious circumstances; one might also inquire of those respondents of environmental cost-benefit questionnaires why they are prepared to recognize 'existence value' of certain 'environmental goods', including living entities; comparative studies with societies that evince a more direct relationship with non-human parts of nature could also be illuminating.

This way of proceeding has several advantages over ethical approaches which seek to establish a more abstract calculus of the 'intrinsic value' of non-human entities: in identifying what people already do care about one avoids the need for the sort of moral exhortation that is hard to justify and hard to make effective; it also usefully shifts attention onto the social factors that inhibit its full expression, since if care does not translate into action this is not necessarily due to lack of motivation to act, but to lack of channels for action. If society were organised differently, the care could manifest itself. Small illustrations of this are provided by those societies whose citizens, when provided the means of ecologically benign practices – like recycling or avoiding motor car usage in towns – actually take advantage of them. In focusing on the social and cultural dimensions of the problem one avoids playing – usually ineffectively – on individual conscience and instead focuses on collectively agreed norms. With such norms in place, their formalisation into something akin to constitutional principles becomes a real possibility.

Conclusion

The final emphasis on collectively agreed norms is important. *Care*, as particularistic, can be arbitrary in its attachments: if its greater openness and considerateness are to guide policy or legislation, therefore, this would only be possible or legitimate if people agreed to the *justice* of the obligations thereby generated. The proposition that consensus may move towards a greater concern for non-human interests is only a hypothesis, not a foregone conclusion. I have suggested, though, that the hypothesis is not self-evidently counterfactual and have very roughly indicated some sorts of ways in which it might be tested. One final reason for cautious optimism on behalf of those already committed to norms of interspecies solidarity derives from the nature of rational human ethics itself. If legitimate consensus is moved by the force of the better argument, then those who are partisan in favour of interspecies solidarity can legitimately appeal to a non-formal feature of what counts as a better argument within ethics, namely, that considerations tending to enhance virtuous conduct should take priority over considerations that would diminish it. Thus once reasons to care are articulated, the onus on any reasonable person, I think, is to produce not just objections, but *better* reasons not to care.

Notes

1 The most influential being that of John Rawls (1971). The problems I go on to discuss also bear on the discourse ethical variant of proceduralism as developed by Jürgen Habermas.
2 Obviously many everyday actions are assumed to be morally indifferent and this, I take it, is rational and justified if they are components of habits and practices that have themselves at some time been critically scrutinised. The burden of my argument is not to insist that literally every proposed action be scrutinised, only those which are part of a practice or habit which itself has been called into question.
3 There is still the question of how particular human and non-human interests relate, but I come to this in the next section. The present discussion is focused on justice, as *essentially* about generalisable principles.

References

Benton, T. (1993), *Natural Relations: ecology, animal rights and social justice*, Verso: London.
Dickens, P. (1996), *Reconstructing Nature: alienation, emancipation and the division of labour*, Routledge: London.
Gilligan, C. (1982), *In a Different Voice: psychological theory and women's development*, Harvard University Press: Cambridge, MA.
Hayward, T. (1994), 'Kant and the Moral Considerability of Non-Rational Beings' in R. Attfield and A. Belsey (eds), *Philosophy and the Natural Environment*, Cambridge University Press.
Hayward, T. (1995), *Ecological Thought: an introduction*, Polity Press, Cambridge.

Hayward, T. (1997), 'Anthropocentrism: a misunderstood problem', *Environmental Values*, Vol. 6, No. 1, pp. 49–63.

Kant, I. (1963), *Lectures on Ethics*, trans. L. Infield, Harper and Row: New York.

Kant, I. (1964), *The Doctrine of Virtue*, trans. M. Gregor, Harper and Row: New York.

Okin, S. (1990), 'Reason and Feeling in Thinking about Justice' in C. Sunstein (ed.), *Feminism and Political Theory*, University of Chicago Press: Chicago and London.

Rawls, J. (1971), *A Theory of Justice*, Oxford University Press.

World Commission on Environment and Development (1987), *Our Common Future*, Oxford University Press.

5 Discounting, Jamieson's trilemma and representing the future

Robin Attfield

The literature on the nature and extent of obligations to future people abounds with paradoxes. Where discounting involves multiplying the value of future benefits and costs by an annual factor which is normally less than one, conventional discounting appears a practical necessity and at the same time an outrageous infringement of impartiality. And matters do not improve if, to avert our intuitive distaste at its long-term effects, discounting is applied to the near future only and not to the further future. Yet attempts to reject the resulting trilemma are often equally disastrous, or so I shall be arguing. I shall also be arguing that discrimination against future people is likely to continue until they are represented when significant decisions are being made.

In the first section, the trilemma is introduced and proposed solutions to it are found wanting. In the second section, after a review of the arguments in favour of discounting, I tentatively conclude that, except where there are special reasons to the contrary, future benefits and losses should receive the same weight as if they were situated in the present. The third section tackles problems for this stance arising from the trilemma and thereby reinforces the earlier conclusions. The fourth and final section introduces proposals for institutional reforms involving the representation of future people in present decision-making.

1

Dale Jamieson sparked off these reflections with his recent paper 'Future Generations', in which the trilemma of my title is posed.[1] Besides presenting the options which constitute the trilemma, Jamieson also presents apparently conclusive objections to all three.

In the course of presenting the conventional approach, Jamieson indicates that there is room to discuss exactly what the discount rate discounts and I shall be returning to this question. Assuming, however, that what is at issue clearly enough consists in the linear discounting of benefits and costs, Jamieson presents criticisms of such discounting, which include the way in which it seems to go too far in assigning quite small values to huge future costs and thus making future interests virtually disappear. To this intuitive objection he adds the related moral objection that in this way discounting disregards the rights and interests of future people (which are hidden by the phrase 'future costs') and also the psychological objection that people support discounting out of weakness of imagination, in that future people do not seem real to us in the present (Jamieson, 1995, pp. 2–6).

The obvious alternative to consider is that the pure discount rate should be zero. This position is ascribed to Derek Parfit, but Parfit is recognised to allow discounting as 'a crude rule of thumb' (ibid., p. 6). (My impression is that Parfit goes further, for he declares that he has no quarrel with a Social Discount Rate (SDR) 'applied to benefits and losses measured in monetary terms, on the assumption that there will be inflation'. What he is against is the application of an SDR to benefits and losses, 'measured at the size they will have when they occur', such as the actual utility which future people will enjoy, or (again) future deaths (Parfit, 1984, p. 480).)[2]

While the criticisms of conventional discounting seem to be reasons for a zero SDR, Jamieson finds insuperable problems for this stance in the vast numbers of the future people likely to succeed us. One aspect of the problem lies in the probably undiscoverable nature of their preferences. Still worse is the likelihood that present preferences would be outvoted by future preferences on many, if not most, issues affecting the environment and that most actions involving change would thus turn out to be wrong. Indeed this consideration could inhibit action in every generation. Further, the huge gulf between the extent of obligations to future people on the zero SDR account and the unconcern expressed towards future generations in our present behaviour suggests that we are actually motivationally incapable of compliance with these obligations and therefore cannot be expected to comply (Jamieson, 1995, pp. 6–8).

The third horn of Jamieson's trilemma seeks to tackle the impact of discounting for the distant future of half a century or more away and consists in the proposal of some kind of non-linear discount rate. From some juncture such as 50 years ahead of the present, the discount rate either diminishes or becomes zero; in this way, our discomfort with conventional linear discounting is partially tranquillised, without the difficulties inherent in a linear zero rate being incurred. But this stance does not even deserve the name of 'theory'; it has no ground in principle and explains nothing. To adopt this stance is to cave in before the problems, not to solve them (ibid., p. 8). It may save time to convey that I fully concur with this appraisal.

Now, ways out of dilemmas and trilemmas include defending one of the horns, but all such resorts are no longer available to Jamieson. The other options which he considers consist in disabling the trilemma, in dissolving it, or in reframing and assimilating the problem to one that is better understood (ibid., p. 8). These options are now considered in turn.

Jamieson's disabling strategy takes the form of distinguishing between resources and non-resources. The view now seems possible that the value of resources can be discounted, but not the value of non-resources. Under resources are included both money and anything else the value of which can be exhaustively expressed in monetary terms. A reason for discounting resources might be the reason given by Parfit, namely that they are liable to inflation. Jamieson's examples of non-resources include rights, interests and wellbeing, which, he suggests, should not be discounted because there is no plausible monetary equivalent to them and because they cannot be invested or produce more in the future. Nor can the value of irreversible conditions like the existence of a species be exhaustively monetised either. The dilemma is disabled by applying conventional discounting to resources and a zero rate to non-resources, thus adapting the domains of each in a complementary manner (although Jamieson explains the disabling strategy differently (ibid., p. 9)).

Reasons for discounting resources and not discounting non-resources will make a later appearance. For his part, Jamieson rejects this strategy, partly because the distinction between resources and non-resources is contestable and partly because, right across the non-discountable realm of non-resources, the preferences of future people will continue to swamp those of present people.

There is no need to follow the detail of Jamieson's discussion of *dissolving* the problem. Here he discusses the theory that nothing is owed to future generations since their very existence depends on present policies and so they are not harmed by the choice of one policy rather than another. But Jamieson eventually adopts Parfit's view that, while our obligations are not owed to determinate individuals, we still have obligations with regard to whoever there will be and so the original problem remains undissolved (ibid., pp. 10ff.). Since few if any determinate individuals are affected for better or for worse whichever policies we adopt, the grounds for arguing from the rights of people of the further future seem to me tenuous, though Jamieson would probably resist this conclusion. This conclusion would allow rights, as opposed to interests, to be dropped for present purposes from the class of non-resources and from the moral argument against discounting costs and benefits of the further future.

Jamieson's discussion of *reframing* the problem consists in the suggestion that discounting is relevant to the reasoning of a single person reflecting on her own future, but inappropriate to the decisions of one group of agents which affect a quite different group (ibid., p. 12). Here the reply would be that this is not at all obviously true, particularly where resources with a monetary value are in question.

However, Jamieson's point is strengthened when he presents the strategy of assimilating the problem to one better understood. Here he points out that most future people will not be our descendants, but other people's. So discounting is largely irrelevant because duties to future generations are duties to people across political boundaries and mostly not to the future members of our own political community. As such they are to be assimilated to duties to our contemporaries. Besides, the facts that contemporaries exist independently of present actions, and that future people do not, have no moral relevance (ibid., pp. 12 ff.). Jamieson claims that this assimilation strategy constitutes progress, since questions about the nature and basis of our obligations to contemporaries are less intractable than the corresponding questions with respect to future generations and discounting (ibid., p. 14).

Now it may be granted that there is some analogy between obligations to contemporaries and to future people. Thus, plausibly, both contemporaries and future people either have or will have certain basic human needs and these needs could supply the grounds of our obligations in both cases. Others might here appeal to the fact that they all either have or will have preferences, which might also supply such a basis; but the unpredictability of future as compared with present preferences makes this, in my view, a source of disanalogy.

But whatever the analogies between obligations to contemporaries and to future people, disanalogies remain, corresponding to the various arguments in favour of discounting future (as opposed to present) costs and benefits. Since Jamieson considers discounting plausible in matters of resources and it is on resources that many issues of possible obligations to future generations focus, he is in no position to claim that assimilating the question of obligations to future generations to the question of obligations to contemporaries makes it unnecessary to form some view about discounting in answer to the question about future-related obligations, even though this involves addressing his trilemma all over again. In any case the arguments in favour of discounting need to be reviewed.

2

These arguments are discussed by Derek Parfit (1983; 1984, pp. 480–86) and (in most cases) by John Broome (1992, pp. 52–112) and also by David Pearce and his fellow-authors of *Blueprint for a Green Economy* (1989, pp. 132–52). In reviewing Parfit's discussion we should bear in mind that he has no objection to an SDR applied to benefits and losses measured in monetary terms.

Parfit first considers the argument from democracy, in circumstances where the electorate cares little about the further future. He points out that, whether or not this argument is relevant to the question of what the government should do, it is not relevant to the question of whether we as a community ought to be diminishingly concerned about impacts as they recede into the future (Parfit, 1984, pp. 480ff.).

The next argument to be considered is that from probability; we should discount more remote effects because they are less likely to occur. He replies that this argument gives no support to an across-the-board SDR; it supports nothing more than discounting predictions which are more likely to be false and there need be no proportional correlation between this and remoteness in time. Remote bad consequences are no less important for being distantly located in the future (ibid., pp. 481 ff.). These appraisals are effectively endorsed by David Pearce and his colleagues (1989, pp. 139ff.). This argument, then, supports nothing more than selective discounting based on probability and does not support linear or exponential discounting at all. Incidentally the same may be said for the argument that future benefits and losses should be discounted because there may be no future people; it is relevant only in proportion to the probability of human history ending by the times in question.

Next to be considered is the argument from opportunity costs. The delaying of certain benefits, which a zero SDR would encourage, involves foregoing the returns which present investments can generate; returns which could themselves be reinvested profitably through the period of the delay. Opportunity costs are the losses involved when benefits are deferred and are often held to justify a positive SDR. However, some benefits are not reinvested but consumed and where this is done, the alternative option consisting in a deferment of benefits would involve no opportunity costs and no discounting is justified. Again, the fact that future costs could be compensated through investing a smaller sum in the present does not justify discounting all future costs, as many will not be compensated. Such arguments at most justify discounting in cases where compensatory action is taken before a predicted evil, or where investments are made to generate future benefits, and then only in periods where investments really do generate interest (Parfit, 1984, pp. 482–4). Pearce and his team (1989, pp. 141–3) reject this appraisal, but only, apparently, on the basis that since the possibility of compensation is held to be sufficient in cost/benefit analysis, the same should be held about discounting. Yet here it would be equally tenable to conclude that cost/benefit analysis ought to be modified. Thus Parfit rightly concludes that the argument from opportunity costs, like that from probability, fails to justify across-the-board discounting, but may justify a selective positive SDR in appropriate circumstances.

When Parfit turns to the argument that our successors will be better off, he sweeps up into it an argument from inflation and the diminishing marginal utility of money. To the argument concerning the comparative wealth of our successors, he sensibly replies that many of them may not be wealthier or in other ways better off than our contemporaries and hence that they may be ineligible to suffer from the diminishing marginal utility of money (Parfit, 1984, p. 484). Again, Pearce's team largely agree (1989, pp. 140ff.). Here, however, Parfit's readers need to remember that he does regard predictable inflation as a good ground for the discounting of monetary benefits and losses. But presumably he could comment that the argument from predictable inflation is no better than the argument from opportunity costs, at least with regard to applying only to periods and domains of history where inflation is predictable. A zero inflation rate would ruin this argument and so would the unpredictable surges of inflation characteristic of the Weimar Republic or the USA of the period just after the Wall Street crash.

The argument from excessive sacrifice claims that without a positive SDR, small but long-lasting future benefits could require large and unacceptable sacrifices in the present. Here Parfit suggests that the real objection is not to belief in the importance of future benefits but to a very unequal distribution of costs between generations. As Parfit adds, utilitarians would claim that this is not implied by the objective of maximum net benefits over time. If, however, such a distribution really would be involved, our resort in order to forestall it need not and should not be the adoption of a positive SDR. Rather we should, he suggests, adopt a second principle which precludes an unfair sharing of benefits and burdens between generations, a principle which would actually be compatible with a zero SDR (Parfit, 1984, pp. 484ff.). Broome (1992, p. 106) accepts this suggestion, as do I. Here it could be added that the high acceptance-utility of this second principle makes it likely that it would in any case be derivable from the maximising principle.

The final argument to be considered by Parfit is the argument from special relations. Common sense morality authorises us to give priority to the interests of our children, our friends and our fellow-citizens, as opposed to those of strangers; hence the people of the further future should receive a lower priority. Without rejecting the claim about priority, Parfit replies that some weight should still be given to the interests of stranger,s and analogously to the further future, and that a positive SDR fails to do this. Also the priority argument cannot authorise the infliction of grave harms or deaths. Thus the argument from special relations, like the others, fails to justify across-the-board discounting, a conclusion which Broome again finds convincing (Parfit, 1984, pp. 480 ff.; Broome, 1992, p. 108). Here it may be added that the kind of consequentialism which upholds optimistic social practices can explain both the special obligations to relations and fellow-citizens and the importance of not giving these obligations unqualified priority.

Broome also discusses Pearce's arguments from the possible bad environmental effects of lowering the SDR, through the encouragement of faster economic growth. While these effects could well be likely in practice, this could not possibly be because of greater weight being given to future interests. The explanation would rather be some failure in the calculations. Thus cost/benefit analysis continues to ignore environmental externalities. If the SDR were lowered and externalities were taken adequately into account, people living in the future would be likely to benefit. This argument, then, only supports a continuation of discounting in a skewed economy (Broome, 1992, pp. 101 ff.; cf. Pearce et al., 1989, pp. 144–51).

For his part, Parfit adds that where the arguments which he has considered do not apply, we ought to be equally concerned about the predictable effects of our actions, however distant in time they may be (Parfit, 1984, p. 486). Thus pure time-preference is irrational and where distant serious harms are predictable, they count just as strongly as like harms in the present. Broome too (1992, p. 108) finds in favour of a discount rate of nought for pure harms and benefits, though he does not rule out the introduction of a modified interest-rate where monetary measures are applicable (ibid., pp. 91 ff.).[3]

But all this strongly suggests that unqualified conventional discounting is unjustifiable; that sheer time-preference, supposedly the central justification of discounting, is no justification at all; and that, other things being equal, impartiality between times and between generations is morally mandatory, at least where serious interests are at stake. In particular circumstances, special discounts may be in place for uncertainty, opportunity costs or predictable productivity and inflation; and principles concerning a fair sharing of burdens between generations and concerning special obligations may be in place in cases where unfair burdens or neglect of recognised ties would otherwise arise. But in circumstances where such factors are either absent or insignificant, for example where serious harms or deaths are predictable in the further future and are significant enough to dwarf these factors into comparative triviality, future benefits and losses should receive the same weight as if they were situated in the present.

3

Yet Jamieson's arguments may here return to haunt us. Does this tentative conclusion simply reinstate the distinction between resources and non-resources which he finds objectionable? And does it fall foul of his objection about the weight of future people's preferences?

Here are some reasons why the distinction between resources and non-resources may seem to be reappearing. As was previously mentioned, under 'resources' Jamieson includes both money and anything else the value of which can be exhaustively expressed in monetary terms. But these are the very items amenable to opportunity costs, or predictable productivity, or inflation, some of the areas where the tentative conclusion would admit of discounting, and of exponential discounting at that. By contrast, predictable serious harms and deaths, central cases of the inappropriateness of discounting, clearly count as non-resources. Thus the tentative conclusion apparently comes dangerously close to an option rejected by Jamieson. On the other hand, the closeness may be no more than apparent, for the tentative conclusion does not endorse discounting even for items which can be monetised, in circumstances where profits or inflation are unpredictable or where opportunity costs are illusory.

This is a salutary moment to return to Jamieson's objections to the distinction between resources and non-resources and to the suggestion that only the former are to be discounted. One objection is that there is no clear distinction to be drawn; and this claim would be supported by adherents of cost-benefit analysis (such as Pearce's team) who believe that a monetary value can and must be assigned to all benefits and losses, death included. For this belief they have some good grounds, for they are eager to include as many interests as possible in cost/benefit analysis, externalities included, and understandably so, as in no other way can all impacts on interests be included when new initiatives are being costed.

Admittedly one way of costing (for example) large civil engineering projects includes assigning some large but finite value to the predictable deaths of construction workers arising from accidents during the process of construction; and admittedly once a monetary value is given to a death, future costings must consistently adhere to the same valuation for other comparable deaths. Sometimes, inbuilt costings will even approximate to our intuitive judgements about priorities. However, the burden of Jamieson's, Parfit's and Broome's arguments was that applying an SDR to the value of predictable serious harms and deaths of the distant future produces ridiculous results and shows that discounting is inappropriate in such cases. Whether this is because deaths and serious harms cannot be monetised or because discounting is inapplicable to any costings thus generated does not need to be decided here. Serious harms and deaths must in any case count as non-resources, since they do not belong to the realm where discounting has been found appropriate.

Jamieson's objection, however, could be more troublesome in the area of interests other than the avoidance of serious harms and of death. Interests such as the retention or loss of a scenic view or of freedom from noise or of landmarks which confer a sense of identity, are probably included among his examples of non-resources, but are given monetary values by economists at least as readily as injuries, illnesses, incapacitation or death. However, to apply an SDR to interests of the further future such as these is likely to produce results as ludicrous as those of applying an SDR to future deaths; and from this I conclude that they should not be included in cost-benefit analyses, but given special attention among factors which cannot satisfactorily be costed. In Jamieson's terminology, this makes them non-resources, even though the suggestion that they might somehow have been resources instead verges on the incoherent. At all events, his objection about the possibility of giving monetary value to apparent non-resources turns out not to generate substantial problems.

Jamieson's other objection concerns the possibility of present plans being outvoted by future preferences, across the realm immune from discounting. This objection was also raised against the view that the SDR rate should be zero, where it appeared alongside further objections, including the objection that the preferences of future people are unpredictable. Clearly these objections need to be considered together, since their mutual consistency is questionable.

Now one of Jamieson's claims is that if future preferences are hard to discover but capable of being discovered, then apparently we ought to spend more researching the related interests of future people than in promoting the welfare of present people; but this conclusion borders on the absurd (Jameison, 1995, p. 6). So it certainly does; but it is only reached because of two dubious assumptions. One of these is that the preferences of people of more than a few months into the future are discoverable in the present at all; but this is highly implausible. A second is that interests (or possibly needs) are a function of preferences. I have argued elsewhere against this assumption.[4]

But if either interests are not a function of preferences or future preferences are mostly undiscoverable, or both, then the case for basing present decision-making on future preferences disappears. And if so, Jamieson's next objection, about the perennial likelihood of the present being outvoted by the future, collapses, at least insofar as it concerns future preferences.

Future needs and interests are a different matter, since we can plausibly form some understanding of these on the basis of the common humanity which we share with future people. Indeed a parallel objection could possibly be mounted on the basis of these needs and interests. But maybe we often pay too little heed to these needs and interests. There would be an upper limit to the diversion of resources away from the present to cater for these needs and interests, if we accept Parfit's and Broome's suggestion of a principle prohibiting any generation's having to shoulder excessive burdens or avoidably receiving a severely unequal (or a minute) share of resources; and this could prove enough to prevent decision-making in the present being reduced to paralysis, as suggested by Jamieson, particularly if an exponential SDR should often be applied to resources. But the importance of taking these needs or interests into account suggests that some better form of institutional provision should be made for doing so and to this we shall return.

Before such provision is considered, Jamieson's remaining objection to a zero SDR needs to be considered. This is the objection that the gulf between the obligations which this would generate and the unconcern expressed in current behaviour suggests that we are motivationally incapable of compliance with these obligations. Now this might well be true if our obligations extended to responding to all the preferences of all future generations; but, as has been seen, they do not and could not. Once again, a parallel argument could be mounted by Jamieson concerning the gulf between present obligations responding to future needs and people's uncaring present behaviour and the conclusion might again be proposed that, since 'ought' presupposes 'can', it cannot be the case that we ought to respond to these needs. Yet this conclusion would imply that we quite generally have no obligation to make life-and-death differences to future people where present decisions will predictably make impacts of this order and is thus itself implausible. Besides, the supposed gulf is much less vast than that of Jamieson's actual vast-gulf argument and could be tackled with the aid of imagination; tackling it would not, now that preferences are no longer in question, involve prophetic psephology. (Anecdotal evidence was adduced at the conference, suggesting that when citizen groups are consulted, they usually agree that the interests of the distant future ought to be taken much more seriously than happens at present.)

Since Jamieson's general objections to a zero SDR turn out to be inconclusive, they must also be inconclusive with respect to a zero SDR for non-resources. Similarly they must be equally inconclusive if raised as objections to the tentative conclusions reached earlier, which favoured a zero SDR both for what Jamieson calls 'non-resources' and also, in certain circumstances, for what he calls 'resources'. Thus these tentative conclusions stand in face of the most ingenious available objections and should be regarded as among the conclusions of this paper.

4

In any case, the feasibility of tackling the gulf between current performance and future-needs-related obligations would be increased by the introduction of improved institutions and we can now return to this issue. In the absence of advocates of future interests, there is a widespread current tendency to average down the demands of future needs and interests. Thus if the above arguments are sound, considerations of uncertainty, opportunity costs and the like, which apply in some cases but not in all, have been generalised sufficiently to support across-the-board discounting, frequently to the detriment of future people. This tendency could be corrected for by the introduction into governmental decision-making bodies of representatives of future needs and interests. These representatives could and should be supported, both at national and international levels, with research teams charged with discovering probable future needs in areas such as supplies of food and energy, the impacts of global warming and the abatement of pollution.

I am not suggesting that these representatives should be given a majority vote, despite the likelihood that future people outnumber present ones. One reason for this is that the foreseen impacts of current action should be discounted for uncertainty, where this is relevant, as it very often is; and though it is certain that present decisions will have future impacts, representatives of the future cannot speak as if certain of the interests they represent and thus cannot safely be given a preponderance of power. Besides, if they were granted a majority of votes, the remaining decision-makers would come to see themselves as representatives of the present to the exclusion of the future, to the detriment of the quality of decisions taken. All decision-makers should take future interests seriously into account. But the prospects of this happening would be increased if a few specific representatives of the future are present and are given a voice.

Pilot schemes could of course be introduced for an experimental period. Since the representatives of future people could not exercise effective power in this period, they could usefully spend some of their energies on advising current government agencies on issues such as the SDR. For while current bodies are all too likely to perpetuate the use of positive, linear and thus exponential SDRs on the basis that they comprise a crude rule of thumb and that no alternative is on offer, bodies on which future interests had a voice would be likely to lay down criteria prohibiting the use of SDRs where distant impacts are predictable, where future profits and inflation are so uncertain as to make any SDR ridiculous, or where future interests which cannot be assigned a monetary value are at stake. While I should welcome some Parfitean statisticians and economists setting to work on specifying such criteria, and an ensuing public debate about making the use of SDR selective rather than global, my guess is that it would take the encouragement of the kind of body which has just been suggested before a self-interested present generation would take these proposals seriously.

Notes

1   The trilemma will also appear in Dale Jamieson, *Philosophy Down to Earth: Science, Values and Environmental Change* (in preparation).
2   At p. 486, Parfit adds that the various arguments for an SDR at most justify 'the use of such a rate as a crude rule of thumb'.
3   Broome (1994) has also argued for a discount rate for those commodities which have 'an interest rate of their own'.
4   On the independence of interests from preferences, see Attfield, 1995a; on the nature of needs, see Attfield, 1995b, chapter 5; also Sagoff, 1988.

References

Attfield, R. (1995a), 'Preferences, Health, Interests and Value', *Electronic Journal of Analytic Philosophy*, Vol. 3, No. 2, pp. 7–15.

Attfield, R. (1995b), *Value, Obligation and Meta-Ethics*, Rodopi: Amsterdam and Atlanta.

Broome, J. (1992), *Counting the Cost of Global Warming*, The White Horse Press: Cambridge.

Broome, J. (1994), 'Discounting the Future', *Philosophy and Public Affairs*, Vol. 23, No. 2, pp. 128–56.

Jamieson, D. (1995), 'Future Generations', unpublished paper presented to a Workshop of the Swedish Collegium for Advanced Study in Social Sciences, Friiberghs Herrgård, Sweden, 25–27 August.

Parfit, D. (1983), 'Energy Policy and the Further Future: The Social Discount Rate' in D. MacLean and P.G. Brown (eds), *Energy and the Future*, Rowman & Littlefield: Totowa, NJ.

Parfit, D. (1984), *Reasons and Persons*, Clarendon Press: Oxford.

Pearce, D., Markandya, A. and Barbier, E.B. (1989), *Blueprint for a Green Economy*, Earthscan: London.

Sagoff, M. (1988), *The Economy of the Earth*, Cambridge University Press: Cambridge.

6 The Lockean provisos and the privatisation of nature

Markku Oksanen

Introduction

At the very heart of John Locke's *Second Treatise* is an idea of the best social arrangement regarding control of the natural world: the natural world and its constituents should be owned privately. The right of property allows humans to use the natural plenty they own within certain natural and positive limits for the purposes of their subsistence and their convenience. The legitimacy of this process of privatisation rests, partly, on the satisfaction of another pair of rules called Lockean provisos which require people not to acquire property excessively and not to spoil their property.

I shall focus my attention on the meaning of these provisos and the practical implications arising from different interpretations of them with respect to the functioning of the institution of private ownership. My perspective on this issue is that of environmental ethics and my purpose is accordingly confined to identifying the possible environmentalist elements in the Lockean provisos and assessing their role in Locke and in some of his followers in the fields of environmental economics and politics. Recently Locke's liberal ideas has been re-embodied in the doctrines of *free market environmentalism* (FME). FME argues that survival of our species, preservation of nature, sustainable use of natural resources and other ecological aims are to occur by means of imposing the morality and the policy of property rights thoroughly into practice. How do these elements of Lockean ideology, i.e., the extensive privatisation and the rule that actually delimits it, fit together? How have various writers attempted to solve this matter? And, finally, should some of the resources, like land, remain as non-privatised?

I defend the view that the right of individuals to possess natural resources *should not* be regarded as unlimited in itself but that if the provisos are taken seriously they do not entail private control either in the use of one's property or in the acquisition of unheld resources to such an extent as FME suggests.

I shall commence by considering the nature of the Lockean conception of property in general. I shall then analyse the meaning of the Lockean provisos with especial regard to the issue of privatisation of nature in some recent interpretations of Locke. And, before my concluding remarks, the question of public ownership will be examined.

Four basic elements

There are four key ideas in Locke's theory of property that can be brought out from his most famous chapter 'Of Property': the privatisation of the commons, the use of resources privately, the universality of property rights and, finally, the Lockean provisos. In what follows, I shall briefly discuss these elements in turn.

1   Privatisation

In the beginning, starts the Lockean story based on Biblical exegesis, everything on earth was common between men. Nevertheless, God did not mean everything always to stay in common but the preferred an institutional end-state for all resources is private ownership.

2   The use and the appropriation of resources

John Plamenatz (1963, p. 241) has remarked that Locke rests his theory of ownership on two natural rights: the right to use external goods for the purposes of preserving one's life; and the right to possess the results of one's own work (cf. Simmons, 1992, pp. 242–43). Thus, when God gave the earth to humankind, Locke says that 'it cannot be supposed he meant it should always remain common and uncultivated' (§34)[1] but the 'Industrious and Rational' individual humans can and, in effect, ought to enclose into their private ownership some parts of nature by mixing it with their labour. God did not only permit people to use the natural plenty but 'God and his Reason commanded him to subdue the Earth, i.e. improve it for the benefit of Life, and therein lay out something upon it that was his own, his labour' (§32). So it is a divinely imposed duty for a human individual to start to 'improve' the uncultivated land (minimally, for subsistence purposes) and so, as Olivecrona (1974, p. 228) points out, 'Man was, indeed, under the law "for appropriating"' (cf. Rapaczynski, 1987, p. 190). The improved land becomes its improver's – or his employer's – private property (§28). Nowadays this theory is known as the labour theory of property and the acquisition of property from the commons is called original appropriation.

The labour theory of property makes it difficult to draw any rigid distinction between the original appropriation of resources and their later use. For the appropriation of land always presupposes its use and the use of one's property ensures its remaining in one's possession. It is also clear that through the process of privatisation humans come to alter the face of the earth – a process regarded positively by Locke and his adherents.

3 The universality of property rights

Locke and his successors argue for the universality of property rights and by this they can mean at least the three[2] following ideas: (i) as relating to the possible *property objects* and to the privatisation of nature, it can be understood as a thesis arguing that all the objects that can in principle be owned privately should be, so that the range of privately ownable things should be expanded extensively, towards the morally acceptable extreme (see Epstein, 1994, p. 19; Posner, 1977, p. 29; Tietenberg, 1992, p. 45); (ii) as relating to property holders, it can mean that a system of private ownership should be adopted universally, since only the private property system is that kind of social order which entirely coincides with the natural law and the natural right to property. This right ought be recognised as a morally valid claim which is legitimately realisable, despite the different local modes of systems of ownership as they actually exist; (iii) it can also mean both of these things; that is, everything that is universally defined to be ownable ought to be owned privately in every society regardless of whether or not the actual property institution in that society is of that kind.[3]

4 The Lockean provisos

According to the natural law, people may use and acquire natural resources and have a title to it provided that two conditions have been satisfied:

i   that '... there is enough, and as good left in common for others' (§27); and

ii   that 'as much as any one can make use of to any advantage of life before it spoils; so much he may by his labour fix a Property in. Whatever is beyond this, is more than his share, and belongs to others. Nothing was made by God for Man to spoil or destroy' (§31).

C.B. Macpherson (1962) calls the first stipulation *the sufficiency limitation* and the second *the spoilage limitation*. The sufficiency limitation suggests that what one may legitimately take from the commons by means of labour is what his share is and when one exceeds this, he acts contrary to the valid claim of others to these resources. The spoilage limitation requires that appropriation and use must be non-destructive and non-wasteful; acting otherwise violates the interests of others. Both of the stipulations are related to the management of resources: a plot of land is converted into private property through work and is maintained in one's possession by means which neither spoil nor waste these things. The restrictions regard both those parcels of land and resources that exist as unowned in the state of nature and those already appropriated into private ownership – they provide criteria for morally assessing the measures humans take toward the natural world.

Standing of the provisos

There are several discrepant interpretations of what Locke meant by these limitations and why he stipulated them. Some are based on the idea that Locke did not mean these limitations to hold in a monetary economy, but only in a barter economy. This position is advocated most notably by Macpherson.

According to Macpherson, there is a difference between 'the initial limited right' and the unlimited right and this difference parallels the transition from a barter economy to a monetary economy. Although, Macpherson states, by quoting Locke (§4), that at the beginning of the *Second Treatise* every man had a natural right to property 'within the bounds of the Law of Nature', the limitations disappear as a result of this transition. Macpherson says that 'Locke's astonishing achievement was to base the property right on natural right and natural law, and then to *remove all the natural law limits* from the property right' (Macpherson, 1962, p. 199, italics added). What are these 'natural law limits' that were exceeded after the introduction of money? It seems – although Macpherson does not discuss it – that they cannot include the limit of property right in itself because it requires people not to interfere in each other's business. In this regard, not all natural law limits were removed. So what Macpherson refers to are the duty of charity and the provisos, among other possibilities. Considering the latter, Locke's conception of property use and acquisition appears as essentially unlimited and neither of the provisos is imposed in a monetary economy. The agreement over money made it possible and rational to exchange one's products for unspoilable assets, for money: thus one could accumulate any amount of land and other resources without violating the spoilage limitation because 'gold and silver do not spoil; a man may therefore rightfully accumulate unlimited amounts of it' (ibid., 204). But what about the sufficiency limitation? The absence of restriction with respect to the size of an area of land one can legitimately appropriate implies acceptance of the unequal distribution of property. As much as one succeeds in acquiring by means of labour (or by other secondary means), that much has he abolished the presumptive right of others in relation to these resources. As Macpherson sees it, this was tacitly accepted as a consequence of agreement over money and individuals may acquire property to such an amount that leaves others less to acquire (ibid., p. 210). For these reasons Macpherson characterises Locke's *Treatises of Government* as a manifestation of modern capitalism.

Contrary to Macpherson, several mainstream liberal and libertarian theorists have espoused the provisos as sturdily as they have adhered to private property rights (see Waldron, 1979, p. 319n; Simmons, 1992, p. 288). For them, the Lockean provisos are to be considered, within a liberal framework, as politically significant and historically recognised restrictions on appropriation and use, however tight or loose their meaning is practically. They matter through the transitory periods: from the state of nature to civil society, from a barter economy to a monetary economy. Consider the ecological reasonableness of the provisos against the general Lockean background. From this perspective it may appear that the Lockean limitations are reasonable precepts of environmental policy only when viewed independently of Locke's theory because Locke is 'quite an unabashed philosopher of capitalism: an apologist for mastery over the natural world ...' (Rapaczynski, 1987, p. 117). Though many sympathetic interpreters accept the provisos, they also tend to accept this protocapitalistic impression of Locke. The Lockean provisos can be seen to express anthropocentric principles about the proper use of natural resources and to prohibit people from treating them in a way that violates the interests of other people, now or in the future. Nevertheless, I shall argue that, when we approach Locke's text as it stands, the spoilage limitation does not appear so much as a rule which requires one to abstain from using one's own property but, rather, requiring its use 'to the best advantage of life and convenience' (§26). What is of more interest is the sufficiency limitation and how it can delimit extensive privatisation.

The provisos, nonetheless, have not always been accepted as such, but have been taken selectively or in some modified form. Some authors, particularly Nozick (1974, p. 175), apply the notion of Lockean proviso only to the sufficiency limitation while leaving the omission of the other limitation unexplained. Others, like Waldron (1979, p. 321), emphasise the greater significance of the spoilage limitation, even after the invention of money, than that of the sufficiency limitation. How then should these provisos be understood?

The right use of property

The spoilage proviso can be seen to reflect Locke's conception of ownership, i.e., what an owner may do to and with his property, though in general Locke was not particularly precise in this matter. Tully (1993b, p. 121) stresses the element of exclusiveness in Locke's property conception and says that

'Locke's definition of property does not specify what degree of control one has over the object except that it cannot be taken without consent ...'. On the other hand, it has been asserted that Locke does not define property right in that way but articulates here one of its constituent rights (Simmons, 1992, p. 227). There is a tension between the spoilage proviso and the right freely to determine the use of property. This may partly explain some authors' reluctance to acknowledge and accentuate the spoilage limitation: it seems to qualify the content of ownership by denying, as a minimum condition, a negative right which permits the owner to let her property decay or spoil, but it may also be construed as implying some positive duties on the part of the owner.

Is the spoilage limitation able to bridle acquisitive capitalism so that the unsustainable use of resources is prohibited? Although it is intuitively very appealing to understand the spoilage proviso as amounting to an ecological restriction on the right to property, this has not been a prominent stance in recent debates. Perhaps the reason for this is that although the spoilage proviso could justify such restrictions, it is very improbable that Locke would have approved these restrictions to such extent and in such a way as would be ecologically reasonable.[4]

More probably, what Locke considered right use is: if I let my land lie waste, I lose my property rights in it and it returns to the commons where it can again be freely acquired (Arneson, 1991, p. 50; Paul, 1987, p. 204; Mackie, 1977, p. 176). This can be explained by referring to the linkages between the labour theory of value and appropriation of land: labour marks a boundary for private property (Olivecrona, 1974, pp. 232–3) and when this mark vanishes, property vanishes.[5] In that sense, having property in something implies some positive duties on the owner to take care of it, to use it up, or, most generally, to do anything to it that is the opposite of spoiling and wasting, which would decrease its value below its natural value. The spoilage limitation is connected to the condition of having property in something. Practically speaking, Locke argues for intensive, commercial agriculture (see Tully, 1993a, pp. 161–3).

An interesting question in this context is whether the common property that is not used when possible is incorrectly allowed to spoil.[6] We may reformulate this and ask whether there should be any natural, unaffected areas and unused resources or not. Locke's answer is simply that '[L]and that is left wholly to Nature, that hath no improvement of Pasturage, Tillage, or Planting, is called, as indeed it is, *waste*' (§42). What this passage implies is that Locke has in mind a conception of spoilage which includes not only possession that does not exploit property for developmental purposes or put it under intensive cultivation but also frivolous destruction: the waste of 'the common stock' is wrong (§46).[7] The property rules that discourage productive use are unacceptable and the encouragement to productive use may even require governmental action at times (Horne, 1990, p. 60).[8] So, the spoilage limitation clearly says that nothing should be destroyed or wasted by means of forbearing from its use and this concerns both – though differently – private and common resources. Perhaps, when we define it further, the spoilage limitation does not actually qualify the content of ownership but makes it a condition of owning so that the holding of property is an ongoing process, requiring legitimation all the time. What happens to the unused property is that it is not *expropriated* but it simply *reverts* to the commons because the necessary condition of owning has disappeared. As Locke says:

> [B]ut if either the Grass of his Inclosure rotted on the Ground, or the Fruit of his planting perished without gathering, and laying up, this part of the Earth, notwithstanding his Inclosure, was still to be looked on as Waste, and might be the Possession of any other (§38).

So, if we let nature reconquer an area and reconvert it to the state of wilderness, we give up our title to it. Thus preservation of wilderness by means of privatisation turns out to be a conceptual reductio: one has to use one's property and what the proviso mainly expresses is a requirement of productive use of the resources (see Sagoff, 1988, pp. 174–5; Hargrove, 1989, p. 69; Shrader-Frechette, 1993, p. 217). It appears that if any quite wild, untouched part of nature is to be preserved, it cannot be done on a private basis because the land has to be put to some use.

Does the condition also require that nothing must be spoiled by using it? If so, the conception of the content of ownership is more determined than has often been supposed. The attachment to the notion of productive use – and what this practically means is to remain under the authority of the owner – is often included in modern neo-Lockean conceptions of the content of ownership. For example, the following statement by Richard A. Epstein (1985, p. 123) reflects faithfully this ideology and argues for the justifiability of productive use: 'the normal bundle of property rights contains no priority for land in its natural condition; it regards use, including development, as one of the standard incidents of ownership' (cf. Arneson, 1991, p. 50). The earlier suggestions concerning limits placed on productive use and pursuit of economic growth stands in somewhat stark contrast to Epstein's account.

The right extent of privatisation

The sufficiency proviso gives rise to the question of the right extent of privatisation. The most extreme position in this respect has been taken by Terry Anderson and Donald Leal (1992), who call their position free market environmentalism (FME). This view strictly endorses the universalistic idea that all resources – including the local and global commons: the forests; the seas; the sea beds; the roads; etc. – should be owned privately. They insist on the ecological soundness of privatisation and claim it to be the best institutional arrangement for preventing the occurrence of environmental deterioration and resource depletion. Anderson and Leal pay no explicit attention to the issue of Lockean provisos. At first sight, the sufficiency limitation is at odds with the idea of universal property rights over the issue of proper extension of privatisation, because it requires that not all of the resources must be privatised: the acquisition command obliges people to perform acts of appropriation whereas the proviso dictates that some of the common resources should be left for propertyless people[9] to appropriate.

Nevertheless, many libertarians – other than Anderson and Leal – attempt to establish the link between privatisation and sustainability: the more that is privatised, the better. More exactly, although the proviso could be accepted, applying it strictly to practice has not been seen as obligatory: as Nozick (1974, p. 179) points out '[t]his proviso (almost?) never will come into effect'. In one sense this conclusion is naturally reached; were the proviso interpreted to mean that everyone has to abstain from activities that violate it, it would follow that no one may take any appropriative measures when every single act of appropriation decreases the chances of others. In that sense, the proviso might be seen to promote some kind of primitive lifestyle and be against economic growth – but to reach this conclusion one has to have an exceptionally stringent interpretation of the proviso.

Next I shall introduce two modern arguments, or denials of the stringent understanding of the proviso, in which there is a tendency to interpret it loosely and to diminish its normative weight: (i) the denial that the position of others is significantly worsened through appropriation; and (ii) the denial based on compensating the propertyless people for the present (and for the past) appropriation and exploitation of the commons. My claim is, however, that if the sufficiency proviso is taken seriously, it casts serious doubts upon the legitimacy of *extensive* privatisation.

Not harming others

An example of the denial of making others worse off can be found in Nozick who claims in *Anarchy, State, and Utopia* that Locke's sufficiency proviso should not be interpreted too rigidly. He wants to differentiate between two meanings of being worsened by another's act of appropriation; a stringent and a weaker. According to the stringent requirement two conditions have to be met: (i) that others do not lose their opportunities to improve their situation because of a particular appropriation by another; and (ii) that others do not lose their liberty to use what they could previously use freely. He accepts a weaker condition for legitimate appropriation in which the second condition is eliminated and it is acceptable that others are worse off in the following sense: 'for though person Z can no longer appropriate, there may remain some for him to *use* as before' (Nozick, 1974, p. 176). As Nozick sees it, no act of appropriation is allowed to worsen a person's position with respect to the use, rather than the appropriation, of resources.

Therefore, the satisfaction of the sufficiency limitation is not a necessary condition for legitimate *appropriation* since, according to Nozick, any human individual can procure hitherto unowned resources as long as this does not negatively affect other people's present state of wellbeing and violate Nozick's 'baseline', which refers to a situation where there is something left to be used but not necessarily to appropriate.

In the same context, Nozick presents an interesting thought-experiment, known as the 'zipper-argument', to show that the stringent interpretation of Locke's sufficiency limitation leads to infinite regression, whereby no one may become an owner of anything unowned because every time a person acquires a piece of the commons they leave others in a worse situation than they were before the act of appropriation.

> Consider the first person Z for whom there is not enough and as good left to appropriate. The last person Y to appropriate left Z without his previous liberty to act on an object, and so worsened Z's situation. So Y's appropriation is not allowed under Locke's proviso. Therefore the next to last person X to appropriate left Y in a worse position, for X's act ended permissible appropriation. Therefore X's appropriation wasn't permissible. But then the appropriator two from last, W, ended permissible appropriation and so, since it worsened X's position, W's appropriation wasn't permissible. And so on back to the first person A to appropriate a permanent property right (ibid., p. 176).

Nozick's aspiration is, however, to show how the sufficiency proviso should be understood as being compatible with the attitude of acquisitiveness. In order to justify capitalistic appropriation and the distinction between stringent and weaker interpretation, he makes this very un-Lockean distinction between *use* and *appropriation*. The distinction is un-Lockean because in Locke the appropriation, the use and the possession are conceptually tied to each other and it is not self-evident that the opportunity to use must be prior in importance to the opportunity to appropriate.

Nozick's argument is applied in some other ecological formulations which give priority to privatisation over common ownership. The most notable of these is from Garrett Hardin (1968) who believes that the existence of a common property system – where no restrictive regulation with respect to use is implemented – does not motivate self-interested agents to heed their long term good as they would if the common resources were their private property. Were these resources in private ownership, the intrinsic logic of the private property system would lead 'some to hold back resources from current consumption for future markets' and thus to protect the interests of future persons (Nozick, 1974, p. 14). Were the resources in common ownership, there would be a tendency for them to be overused and destroyed (see also Schmidtz, 1990; Wolf, 1995.)

Nevertheless, two questions arise: does this arrangement work properly, that is, will propertyless people be in a position to have all necessary resources?; does this arrangement violate the prima facie claim of future people – especially those family-lines that have so far been propertyless – to appropriate? Regarding the former question, the answer could be affirmative – at certain times, in certain places and regarding certain ownables – but an obstacle arises: the fundamental principles of FME do not restrict the acquisition of property and therefore it would soon be the case that there would be no common land for future appropriation. If we consider that the proviso requires also that something should remain in common and exists in such an institutional form that allows the propertyless to appropriate, neither Nozick's nor Hardin's argument tells for extensive privatisation. Indeed, what the proviso may require is to reverse the process of privatisation because the allocation of resources into private ownership is a permanent, irrevocable measure which probably leads to the situation where the major proportion of the resources is accumulated by some family-lines.

Furthermore, there is also a reason for non-private control regarding the common parts of nature: the sufficiency proviso says clearly that no one should be too greedy in acquiring private proportions from the common resources. If we assume that a state of resource scarcity prevails, the sufficiency proviso would rather imply some kind of primitivism in lifestyle, rather than development based on consumption and appropriation of resources. Locke might thus take an intermediate position. But to reach such a conclusion one has to disregard the other normative elements in Locke which mitigate the prescriptive significance of the sufficiency proviso. In addition, many libertarians – including Nozick (1974, p. 178) – suggest another argument to deny such an anti-growth implication of the sufficiency proviso: the compensation argument.

Compensating others

The compensation argument justifies acts of appropriation in the context of prevailing scarcity. It answers the question of why we may appropriate and use the commons by referring to beneficial consequences ensuing from this conduct: to appropriate (through labour) is to advance development and to make steps to reach a higher level in the evolution of civilisation. Locke's opinion of enclosures is that they are good for the rest of humanity and those who do the hard work of appropriating are benefactors of humanity.

> [H]e who appropriates land to himself by his labour, does not lessen but increases the common stock of mankind ... And therefore he, that incloses Land and has a greater plenty of the conveniencys of life from ten acres, than he could have from an hundred left to Nature, may truly be said, to give ninety acres to Mankind. For the provisions serving to the support of humane life, produced by one acre of inclosed and cultivated land, are (to speak much within compasse) ten times more, than those, which are yielded by an acre of Land, of an equal richnesse, lyeing waste in common (§37).

So, by means of efficiently exploiting raw materials under a regime of private property, we should not diminish the quality of life of the propertyless but quite the opposite; the invention of money made it possible to invest in technological development, in the creation of infrastructure and in the production of goods (cf. Kavka, 1981, p. 120). This kind of development would be the best way to benefit present and future propertyless. Although the amount, availability and quality of natural resources in common and the general conditions of living would not remain alike and unchanged through the course of time, this advancement could compensate the propertyless for the fact that some family-lines have less left them to appropriate: appropriation is based on labour and labour increases the value in nature, which adjusts the declining chances of appropriation. In addition, when the unowned resources are being used, the situation of others is improved, since it is made possible for them to have new kinds of things. As Ellen Frankel Paul (1987, p. 204) highlights, appropriation is not theft – a violation of others' rights – but it 'constitutes a benefaction upon the rest of mankind'. But does it justify negligence toward the sufficiency limitation?

Seeking an answer to this question calls our attention to the nature of the substituted interest of the propertyless. Let us consider this issue in terms of future generations. What is peculiar in the compensation argument is that it puts forward, explicitly or implicitly, a substantive suggestion about the quality of the interests of future people and alleges that they would prefer a technologically advanced, completely privatised world to one which was less privatised and less technologically advanced. In this respect it resembles the preservationist argument based on certain assumptions about the content of the interests of the future people and about the conditions necessary to satisfy their interests. Therefore, the problem is identical: how may we decide what kind of world future generations would prefer and what kind of interests they will have?

If we accept the sufficiency limitation and if we believe that we may decrease future generations' possibilities – and this concerns particularly the propertyless family-lines – through the acts of appropriation, but we see that the development of technology and the improvement of the land are ways to compensate all this, then we have to give an argument that future people would see their good in the same way as *we* see their good. One thing is sure: we cannot take for granted that they would prefer a technologically developed libertarian utopia to any other option.[10] A theory of human good has to be established and to do this, in a theoretical context which mainly relies on subjectivism (see O'Neill, 1993, pp. 36–8), is not easy. This is not so difficult for Locke, because of his moralism, as for his modern successors. Therefore, the argument from compensation hardly provides the potential weight to justify the process of extensive privatisation.

Opposing extensive privatisation

Locke's attitude to the natural world demonstrates some typical anthropocentric features connecting it to earlier metaphysical views: the earth and all the resources were meant for human use, but not unconditionally. This attitude is exemplified both in the command to appropriate and in the spoilage limitation in the sense that the owner is obliged to accomplish certain developmental tasks with respect to the resources, to himself and to other people. This can be ecologically precarious. How should we, then, understand the sufficiency limitation? Does it entail public control of common resources so as to protect the share of the propertyless posterity? My answer is positive and I claim that the intrinsic logic of Lockean thought leads to a requirement that there should be rules to govern access to and use of common property resources as they have existed so far and that these rules imply the justifiability of some kind of public control and monitoring of those resources to ensure that something remains. Consider fishing and fisheries. It is not allowed, according to the sufficiency limitation, to catch all the fish in the oceans or to allocate such exhaustive fishing rights to people (e.g., individual transferable quotas) as would leave nothing in common. Something should be maintained outside this arrangement and exist in common. This issue should be seen as distinct from the question of the institutional frames of sustainable use of resources: is the market approach enough – as the advocates of FME are professing – or does it require some other kind of institutional arrangement like those of public monitoring and management or community-based resource management? This question is an empirical one, to be addressed in comparison with other arrangements (see Becker and Ostrom, 1995).

Regardless of whether a market-based or any other solution leading to sustainable development is possible, the sufficiency proviso does not insist merely that there should always remain some resources over time, but, more specifically, there should always remain non-exploited and unlaboured common property resources.[11] Because of this feature, the sufficiency proviso has implications which moderate privatisation. Firstly, it implies the justification for public control over some common resources because it is not possible on a private basis within the Lockean framework: it is needed to ensure the chance to make original appropriations by the propertyless. Secondly, and perhaps more importantly, it entails patterned rules of just property acquisition. Because there are limits to the amount of resources that each individual, as a member of a generation, is allowed to acquire and use, it follows that the institutional arrangement that a society must comply with inevitably consists of rules of justice that are essentially patterned and not the arbitrary, contingent historically modified sets of entitlements that are regularly upheld by libertarians.[12] If the proviso tells us the amount each generation may use and appropriate, then it necessarily tells us the amount each individual is entitled to have. (How to determine standards of the just distribution of resources cannot be dealt with here.)

Concluding remarks

To sum up Locke's fundamental ideas, we should emphasise that people have a prima facie right, if not a duty, to use resources and to take them into their possession, but not in an arbitrary way. The spoilage limitation is stipulated to enhance the productive use of resources. It is difficult to see how it could be applied, as a general, supreme norm, to advance comprehensively ecological values. The sufficiency proviso, when it is considered to hold for all time, does not speak for the idea of privatising the commons extensively. Rather, it calls for public control of common resources to ensure that something really will remain.

The different elements – the privatisation, the appropriation, the use of the resources and the provisos – in Lockean theory collide, but this does not necessarily signal its insufficiency; rather, the ethical problem focused upon here, and as it was identified in Locke's *Second Treatise*, is also a paradigmatic moral problem transcending all systems of ownership: what are the rights of the propertyless, how much and in what way may we consume and appropriate resources and how should we compensate the propertyless for using and appropriating of resources? So, in applying the Lockean provisos in nature conservation – so that the use of natural resources ought to be limited and something be left for future generations and the propertyless – it has to be made clear what are the links between environmental obligations, property rights and our obligations to the propertyless: the strategy of extensive privatisation does not represent such a policy in a coherent way.

Notes

1   References to Locke's *Second Treatise* will hereafter be referred to by bracketed section numbers.

2   A fourth meaning could be that '"the right to property" implies that everyone ought to have some property' (Schlatter, 1951, p. 245, on Bentham's views).

3   It is disputable in what sense Locke would have subscribed to this thesis. On the one hand, he mentions (in §35) that joint property, property that is common by compact (like commons in England), does not have to be privatised. But, on the other hand, this exception does not extend to the Amerindians and to the commons there because these were identified to be in the state of nature (§49). Therefore the appropriation and the European settlements require no consent by the indigenous people and the colonisation was legitimate (see Tully, 1993a.)

4   Kristin Shrader-Frechette (1993, p. 217) has made a similar claim and supports it by distinguishing between an historical and a contemporary Locke, with different environmental attitudes. The contemporary Locke would argue – against the historical Locke – for extensive ecological restrictions on property rights in land. This proposal is problematic because quite often the ecological restriction delimits, if it does not forbid, intensive, productive land use, whereas Locke would have accepted any treatment of land that tends to increase its value.

5   I find it very unlikely that Locke would have introduced the proviso to condemn waste *merely* because 'he did not like it' (Plamenatz, 1963, p. 243).

6   This issue is related to the sufficiency proviso and it may appear that the provisos are contradictory, one requiring productive use of resources of all sorts and the other requiring leaving something for others. But I think that unappropriated common resources are not wasted because they are left for others, for posterity, to appropriate. In addition, as Rapaczynski (1987, p. 213) emphasises, the spoilage here is people-related, not natural (like in the case of rotting wild berries, if not picked).

7   Historically, Locke's stance, which stresses the role of the work, has been seen as a critical reflection on the idle life of the landowning aristocracy (see Sreenivasan, 1995, p. 17; Simmons, 1992, pp. 286–7).

8   So, when Hargrove (1989, p. 66) writes that '[s]ince property rights are established on an individual basis independent of a social context, Locke's theory of property also provides the foundation for the landowner's claim that society has little or no role in the management of his land, that nobody has the right to tell him what to do with his property' this seems to apply only to the cases where improvement occurs.

9   I shall use the term 'propertyless people' or 'the propertyless' to refer to all groups of people that are not present owners. It includes some members of future generations, although some of them are better off because property transfers according to genealogical lines and others are worse off, similar to those of present propertyless people. Usually when we speak in terms of the interests of future generations we are interested in whether they will have the basic means of living and goods for consumption, but in our context the possessions of each family-line are more important.

10 Compare this to the ethnocentric biases in Locke which were used to justify the expropriation of Amerindian lands (see §30; Tully, 1993a, p. 164).

11 Realistically, the privatisation of common resources would in the first place concern more those common resources that are within national boundaries rather than the global commons. But I space precludes my considering this issue in a more detailed way here.

12 The concepts of patterned principle of distribution and of the end-state principle come from Nozick who defines them as follows: '[L]et us call a principle of distribution *patterned* if it specifies that a distribution is to vary along with some natural dimension, weighted sum of natural dimensions, or lexicographic ordering of natural dimensions. And let us say a distribution is patterned if it accords with some patterned principle' (Nozick, 1974, p. 156). He states unambiguously that 'the Lockean proviso is not an "end-state principle"' (ibid., p. 181). Husain Sarkar (1982, p. 59) says that Nozick's theory of acquisition 'is quietly and deeply supported by end-state principles of justice after all' (cf. Shrader-Frechette, 1993, p. 214).

13 I should like to thank Robin Attfield, Juhani Pietarinen and Juha Räikkä for their comments on the manuscripts as well as the audience of the 'Responsibility, Property and Justice' conference, Durham, for comments on the paper.

References

Anderson, T.L. and Leal, D.R. (1992), *Free Market Environmentalism*, Westview Press: Boulder.

Arneson, R.J. (1991), 'Lockean Self-Ownership: Towards a Demolition', *Political Studies*, Vol. 49, pp. 36–54.

Becker, C.D. and Ostrom, E. (1995), 'Human Ecology and Resource Sustainability: The Importance of Institutional Diversity', *Annual Review of Ecolocy and Systematics*, No. 26, pp. 113–33.

Epstein, R.A. (1985), *Takings: Private Property and the Power of Eminent Domain*, Harvard University Press: Cambridge, Mass.

Epstein, R.A. (1994), 'On the Optimal Mix of Private and Common Property', *Social Philosophy and Policy*, Vol. 11, No. 2, pp. 17–41.

Hardin, G. (1968), 'The Tragedy of the Commons', *Science*, Vol. 162, pp. 1243–48.

Hargrove, E.C. (1989), *Foundation of Environmental Ethics*, Prentice-Hall: Englewood Cliffs, NJ.

Horne, T.A. (1990), *Property Rights and Poverty. Political Argument in Britain, 1605–1838*, The University of North Carolina Press: Chapell Hill.

Kavka, G. (1981), 'The Futurity Problems' in E. Partridge (ed.), *Responsibilities to Future Generations. Environmental Ethics*, Prometheus Books: Buffalo, NY.

Locke, J. (1991), *Two Treatises of Government*, P. Laslett (ed.), Cambridge University Press: Cambridge.

Mackie, J.L. (1977), *Ethics. Inventing Right and Wrong*, Penguin: Harmondsworth.

Macpherson, C.B. (1962), *The Political Theory of Possessive Individualism. Locke to Hobbes*, Oxford University Press: Oxford.

Nozick, R. (1974), *Anarchy, State, and Utopia*, Basil Blackwell: Oxford.

Olivecrona, K. (1974), 'Locke's Theory of Appropriation', *Philosophical Quarterly*, Vol. 24, pp. 220–34.

O'Neill, J. (1993), *Ecology, Policy, and Politics. Human Well-Being and the Natural World*. Routledge: London.

Paul, E.F. (1987), *Private Property and Eminent Domain*, Transaction Books: New Brunswick, NJ.

Plamenatz, J. (1963), *Man and Society. A Critical Examination of Some Important Social and Political Theories from Machiavelli to Marx*, Volume I, Longman: Harlow.

Posner, R.A. (1977), *Economic Analysis of Law*, 2nd edn, Little, Brown and Company: Boston and Toronto.

Rapaczynski, A. (1987), *Nature and Politics. Liberalism in the Philosophies of Hobbes, Locke, and Rousseau*, Cornell University Press: Ithaca.

Sagoff, M. (1988), *The Economy of the Earth. Philosophy, Law, and the Environment*, Cambridge University Press: Cambridge.

Sarkar, H. (1982), 'The Lockean Proviso', *Canadian Journal of Philosophy*, Vol. 12, pp. 47–59.

R. Schlatter (1951), *Private Property. The History of an Idea*, George Allen & Unwin: London.

Schmidtz, D. (1990), 'When is Original Appropriation *Required*?', *The Monist*, Vol. 73, pp. 504–18.

Shrader-Frechette, K. (1993), 'Locke and Limits on Land Ownership', *Journal of the History of Ideas*, Vol. 54, pp. 201–19.

Simmons, A.J. (1992), *The Lockean Theory of Rights*, Princeton University Press: Princeton, N.J.

Sreenivasan, G. (1995), *The Limits of Lockean Rights in Property*, Oxford University Press: Oxford.

Tietenberg, T. (1992), *Environmental and Natural Resource Economics*, 3rd edn, HarperCollins: New York.

Tully, J. (1993a), 'Rediscovering America: the *Two treatises* and Aboriginal Rights' in *An Approach to Political Philosophy: Locke in Contexts*, Cambridge University Press: Cambridge.

Tully, J. (1993b), 'Differences in the Interpretation of Locke on Property' in *An Approach to Political Philosophy: Locke in Contexts*, Cambridge University Press: Cambridge.

Waldron, J. (1979), 'Enough and as Good Left for Others', *Philosophical Quarterly*, Vol. 29, pp. 319–28.

Wolf, C. (1995), 'Contemporary Property Rights, Lockean Provisos, and the Interests of Future Generations', *Ethics*, Vol. 105, pp. 791–818.

7 King Darius and the environmental economist

John O'Neill

1   The problem of market boundaries

What is the source of our environmental problems? Why is there a persistent tendency in modern societies to environmental damage? From within the neoclassical economic theory there is a straightforward answer to those questions: it is because environmental goods and harms are unpriced. They come free. Indeed, Arrow (1984, p. 155) claims this to be a thesis peculiar to neoclassical economics: '[T]he explanation of environmental problems as due to the nonexistence of markets is ... an insight of purely neoclassical origin'. If the source of environmental damage is that preferences for environmental goods are not revealed in market prices, then the solution is to ensure that they are. Two distinct ways of doing so are offered. The first is the direct extension of tradable property rights to environmental goods. The second is the construction of shadow prices for environmental goods, by ascertaining what individuals would pay for them were there a market, which can then enter into a cost-benefit analysis for any proposed projects such that their full benefits and costs can be ascertained. The construction of such prices is carried out either indirectly – by inferring from some proxy good in the market, such as property values, an estimate of a price for environmental goods or by using the costs incurred by individuals to use an environmental amenity to estimate a value – or directly, by the use of contingent valuation, in which monetary values are estimated by asking individuals how much they would be willing to pay for a good or would accept in compensation for its loss in a hypothetical market. The extension of prices to environmental goods so that the 'true' value of preferences for them can be discovered is the road to resolving environmental problems. Thus goes the neoclassical position.

That position runs up against a political and ethical objection that runs in the opposite direction; that our environmental problems have their source not in a failure to apply market norms rigorously enough, but in the very spread of market mechanisms and norms.[1] The source of environmental problems lies in part in the colonisation of markets, not only in real geographical terms across the globe, but also in the introduction of market mechanisms and norms into new spheres of life that previously have been protected from markets. The neoclassical project of attempting to cost all environmental goods in monetary terms becomes an instance of a larger expansion of market boundaries. The proper response is to resist that expansion.[2]

The market can cross ethical boundaries in two different ways. First, items that are considered inappropriate for sale might become directly articles for sale on the market: consider the sale of bodily parts, sexual services, reproductive capacities, votes, political office, the means of salvation and so on. Second, relations, attitudes, forms of evaluation and meanings constitutive of the market might be transferred to other spheres: for example, while university education in the UK is not yet a commodity that is directly bought and sold on a free market, there has been an increasing replication of the language and relations of the market – the treatment of students as consumers, the introduction of course contracts and so on.[3] The two neoclassical responses to environmental problems raises issues on both fronts. Free market solutions to environmental problems that attempt to extend tradable property rights to environmental goods raise questions of market boundaries of the first kind. The practice of economic valuation for the purpose of cost-benefit analysis raises concerns about market expansion in its second form: it represents an incursion of market based norms and modes of arriving at choices into non-market spheres where they are inappropriate. It is this second form of erosion that I consider here.[4]

The problems of market boundaries it raises are graphically illustrated in the resistance of respondents in contingent valuation surveys to expressing their commitments in monetary terms. This can be expressed in direct protest bids, by refusal to engage in the survey or accept a price. However, even those who do put in a bid may not be happy to do so, a point that comes out in qualitative studies of contingent valuation. My aim in this paper is to defend the view that there is something misconceived about the neoclassical project by reflecting on the detail of the responses made to such surveys. I do so by focusing on two valuation studies.

2   Two surveys

My first valuation study is for a wildlife enhancement scheme (WES) on Pevensey Levels in East Sussex undertaken on behalf of English Nature. The study has the virtue of having been paralleled by a focus group study examining in detail individuals' responses to the contingent valuation. Those responses will be the occasion for the reflections in this paper (Burgess, Clark and Harrison, 1995). I refer to both below. The Levels themselves are a large wet grassland system designated a site of special scientific interest (SSSI) with important plants and invertebrates associated especially with the ditch systems. The respondents, drawn from residents, visitors and non-visitors to the Levels, were asked first if they were willing to continue to pay taxes towards WES in the Pevensey Levels. Those who were willing were then asked the maximum they would be willing to pay more for the WES – twice as much, three times as much and so on. As with any such study, some respondents protested, refusing to engage in the valuation project. These protest bids together with strategic bids, 'wild guesses' and the like are rejected. Legitimate bids are aggregated and, as is often the case, it turns out that the sums come out more or less all right.[5] Environmental benefits outstrip costs. And this is the virtue of the method for the environmental economist and for a number of environmental organisations. It gets the right results. So why are other environmentalists against the procedure? What is the worry about market boundaries?[6]

I want to start my answer to that question by contrasting this recent survey with another, older, willingness to accept survey of a non-environmental good. I do so in the spirit of historical scholarship. I offer what must be one of the earliest of such surveys. The report comes from Herodotus's histories (3.38).

> When Darius was king of the Persian empire, he summoned the Greeks who were at his court and asked them how much money it would take for them to eat the corpses of their fathers. They responded they would not do it for any price. Afterwards, Darius summoned some Indians called Kallatiai who do eat their parents and asked in the presence of the Greeks … for what price they would agree to cremate their dead fathers. They cried out loudly and told him to keep still (McKirahan, 1994, p. 391).

I start by noting an obvious contrast between Darius's willingness to accept survey and the willingness-to-pay surveys of his modern economic counterpart. Darius's survey aims to elicit protest bids. The story would have been ruined if the Kallatiai had responded by putting in a realistic price. The reason Darius elicits the protests is to reveal the commitments of the individuals involved. In contrast, the modern economist begins by ignoring all protest bids: these, together with strategic response, are laundered out of the responses to leave us with just those bids relevant to a calculation of the welfare benefits and costs of the project (Willis et al., 1995, pp. 61ff.).

Thus set up, the paper might proceed through a neat contrast between the level headedness of Darius and the follies of the modern neoclassical environmental economist and end fairly quickly. However things are not quite so clear cut. The story of Darius as it is told by Herodotus might not work

116

quite so straightforwardly as a first look might suggest. The problem is that I have omitted the ending. Herodotus rounds off his account of the protest bids with the comment (3.38) (McKirahan, op. cit.): '[T]hat is what customs are, and I think Pindar was right when he wrote that custom (nomos) is king of all'. Now, one way of interpreting the tale is that Herodotus aimed to illustrate the variability of standards of conduct according to the customs of a culture against those who held that there existed universal standard grounded in nature (physis).[7] What is horrific for the Greek is the norm for the Kallatiai and vice versa. Custom is king of all.

When the full text is thus interpreted things might look better for the modern economist, for it might be taken precisely to undermine the appeal the critic of contingent valuation makes to the apparent oddity of asking for monetary valuations. Thus, consider Holland's strategy of undermining contingent valuation for environmental goods by pointing to the peculiarity of using the method in other policy settings (Holland, 1995, p. 22):

> [T]he equivalent questions might be: 'how much would you pay to secure the ordination of women priests ...?' or 'how much would you pay to see hanging retained/abolished? Looked at dispassionately these are very *queer* questions, and this seems to be a very *queer* way of dealing with issues of this kind.

Now one response to that line is simply to say, as does Aldred (1995, ch. 1), very well it sounds very queer to us – but then so do a lot of things that are unfamiliar to us. And here Herodotus looks a fine ally for Aldred. The 'queerness' reaction, either by philosophical critics or respondents, does not show there is anything in principle wrong with using monetary values to come to decisions about environmental policy or hanging any more than does the reaction of the Kallatiai show there is something terrible about burning one's dead kin. The queerness reaction simply points to some cultural facts about us and there is no reason for the economist to be particularly concerned about them. If the rational use of resources requires a cultural change in individuals' readiness to employ monetary values in new arenas, then so be it. Our environmental problems clearly do point to the need for cultural change.

Now, there are a number of different lines of response that might be made to this. It might be objected, for example, that the economist cannot have it all ways: one cannot properly appeal to some universal picture of rational economic choice that applies always and everywhere and then call upon a bit of cultural relativism when things get tough. However, that line of response I leave aside here. Rather I want to pursue the thought that there are non-local market boundaries that are being transgressed here and that they ought not to be.

3   Markets and social relationships

I start by taking issue with Herodotus if interpreted in the manner outlined. His appeal to variability in responses in the King Darius example is less convincing than first look suggests. The differences in question, like any other pair of significant differences, depend on the existence of shared understandings. King Darius's elicitation of protests relies on two sets of understandings that are shared by the two groups, the Greeks and the Kallatiai. The first is the existence of special ties to particular others, more specifically to one's dead kin. The two groups express those differences in different ways – one by cremation, the other by eating – but the differences make sense against the background of a notion of the respectful treatment of the dead. One part of cross-cultural interchange where this can issue in understanding lies precisely on the interpretation of the actions of others that makes them not only understandable but expressive of relations and attitude that we share at a deeper level. What in our culture would be shocking, eating your mum and dad, in another turns out to be expressive of a relation of respect to them.

The second shared assumption in the Darius passage is that about market boundaries; that to accept money to treat the person in a way that would display in that culture disrespect is unacceptable. It is not an accident that Darius elicits the different groups' different commitments by attempting to bribe them. It is not just shared understandings between the groups that is at work here. Our understanding of the passage, unaided by complex anthropological footnotes, relies on the fact that these assumptions are shared also by ourselves. That is why the passage retains its power.

Certain kinds of social relation are constituted by particular kinds of shared understanding which are such that they are incompatible with market relations, where these exist. This is in particular part of what makes relations of personal commitment to others what they are. Given what love and friendship are, and given what market exchanges are, one cannot buy love or friendship. To believe one could would be to misunderstand those very relationships. Some social loyalties – for example, to friends and to family – are constituted by a refusal to treat them as commodities that can be bought or sold. To accept a price is an act of betrayal, to offer a price is an act of bribery.[8]

Moreover, some such relationships of commitment to particular others, for all the variety of ways in which these can be expressed, are components of what it is to live a good life. Given the kinds of beings we are, neither angels nor beasts, we need them. Central to the criticism of the spread of market relations into other social spheres is the corruption this entails of social relations that are central to a good human life. The social meanings that are constitutive of markets are such that they are incompatible with other relationships – and for this reason what matters is not just whether an object is sold but the spread to other spheres of the forms of understanding that are constitutive of markets.

How far does this take us along the way to understanding the problems with contingent valuation? The answer might seem to be, as yet, not very far. What we have been talking about is social relationships, not environmental goods and the implications for the latter might look limited. There are clear cut cases in which the economist acts like a latter day Darius but hoping for a different response. Take the attempt to put a price on Coronation Hill, a part of the Kakadu Conservation Zone in Australia, an area which for the Aboriginal population is a sacred site containing their own dead ancestors. Now, well intentioned as the Australian Resource Assessment Commission might have been in asking how much people would be willing to pay to preserve the Conservation Zone and stop mining in the area, the question is inappropriate. As far as the Aboriginal population is concerned, to ask this question would be to ask how much you would be willing to pay for someone to stop desecrating a grave – to which the proper response is simply that they should stop. To enter into monetary discussions is to collude. To ask the alternative question – how much you would be willing to accept in compensation – is to attempt to corrupt the social ties of a community. A decision to conserve should be made, as indeed it ultimately was, directly in terms of the values involved, not via monetary values. However, it might be thought that a case like Kakadu is a peculiar one, for what matters here has more to with a particular set of social relationships embodied in a site than the natural environment. Most environmental matters are more mundane, Pevensey Levels, cases which are not like that.

This response is mistaken. The view that Pevensey Levels type problems are not about social relationships depends on an unhelpful contrast that the Green movement has inherited from New World wilderness cousins, that what is of value is simply nature considered in separation from humans. A large number of environmental issues are about the kinds of social relations in which we engage and the communities in which we live. The environment matters as a place that embodies particular relations of the past and through which relations to the future are expressed. Consider the discussions of the responses to the Pevensey Levels survey that emerged in the in-depth group discussions of the local residents. The value of the Levels lies in part in the very particular history of social relations it embodies and its destruction matters in virtue of disturbing physically embodied memories. The comment of one of residents, Kate, captures a fairly common reaction to the destruction of local landscapes and habitats.

It's recalling memories of my childhood down on the beach at the Crumbles, we used to a spend a lot of time down there with my dad, sea fishing, and it was all the plants and stuff he was just mentioning. I'm thinking now when I drive through there in the morning that there's *none left at all*. It's quite sad that these things have gone (Burgess et al., 1995, p. 34).

Correspondingly, the environment is expressive of social relations between generations. It embodies in particular places our relation to the past and future of communities to which we belong.[9] And it is that which in part activates the protests to the demand to express concern for nature in monetary terms, including protests from some who may have actually responded 'legitimately' to the survey. Typical is the following response of one respondent, who did not actually put in a protest bid:

> it's a totally disgusting idea, putting a price on nature. You can't put a price on the environment. You can't put a price on what you're going to leave for you children's children ... It's a heritage. It's not an open cattle market (ibid., p. 44).

The point here is that an environment matters because it expresses a particular set of relations to one's children that would be betrayed were a price upon it accepted. The treatment of the natural world is expressive of one's attitude to those who will follow you.

The idea that price is not appropriate here is also articulated in terms of property rights involved.[10] When it comes to the environment in relations between generations the issue of price is rejected because as one resident puts it, 'it's not ours to sell' (ibid.). The point here is that issues of buying and selling only arise if one assumes one has rights that can be alienated. If one understands anyone's relations to environmental goods over time to be those of a usufructuary,[11] a person with use rights but not rights to alienate the good, then the question of either willingness-to-pay or to accept does not arise in the first place. If the environment belongs to your children its not yours to buy or sell. The point here is, I take it, an ethical one, not a legal one. Clearly, one can in fact buy and sell property – the point is that to sell and leave your children with nothing is to betray their claims on a good one has within one's power. As a resident in Pevensey Levels study remarks its 'not just for *you*. It's *your* children and *your* children, *their* heritage' (ibid.). Given that the goods in question are such that only use rights are morally legitimate, the question of payment does not arise.

Particular places and environments are constitutive of a set of social relations over time and markets are properly seen as corrosive of those relationships. Neoclassical economics fails here because it assumes that price is simply a neutral 'measuring rod'[12] of the utility a person expects to receive from an object. It is blind to the social meaning of acts of exchange, the ways in which social relations of particular kinds require the refusal of market norms.[13] The problems of boundaries arise, however, not just with social relations. They occur where concepts of loyalty and betrayal, of commitment and corruption apply. And these apply not just to social relations. They are built into the concept of ethical values. You can betray your values as you can

120

betray other individuals. You can corrupt values by asking for a price. To allow oneself to take prices in that way would be to betray one's values, to allow oneself to be 'bought' by one's opponents. It is because disputes about the environment are not simply about resources, but about values and valued entities – about place, landscape, appreciation of the natural world, about the ways of life in which concern for these are central components – that respondents refuse willingness-to-pay exercises for environmental goods. They display value commitments via just such refusals. The existence of protest bids shows individuals to have a commitment to certain goods which would be betrayed if prices were accepted. Such commitments place ethical boundaries around markets that include environmental goods.

4   Problems of procedure

One of the most interesting and, for me at least, unexpected responses to the contingent valuation question on Pevensey was the following:

Sally: But I'll be quite honest, when I was asked that question, because of the financial situation my husband and myself are in, my answer was 'no, I couldn't afford to give *anything* extra because at the moment we're stretched to our limits. Because of losing jobs and that sort of thing. So, I then missed the bidding, which I was quite pleased about.

Susan: But the trouble is that would be, that could be misinterpreted couldn't it?

Sally: Absolutely. Absolutely. That's right.

Susan: As if you don't want it, you don't care ... But I think, when we filled in that form I feel quite strongly that if we'd known a bit more, and the influence that perhaps the questions and our answers could have. Like when you said you couldn't afford to pay more because of your particular circumstance and if there was no facility to say that, they might get the wrong impression. And that, that fills me with horror to think we all might have been unemployed and we all couldn't have afforded to pay even if we wanted to.

David: That's right. That crossed my mind ...

....

Susan: ... We might *all* have had a reason why we at this moment could not think in terms of paying and that, that bothers me. That bothers me. If it meant the future of the Levels, depended on our particular circumstances (ibid., p. 43).

An initial point to notice about this extract is that at one level this should be of no worry at all to the neoclassical economist. The knowledge that Susan wants is that which would allow her to bid strategically. She wants a particular end – the protection of the Levels – and wants to know what is the best way to bid to get that result. That kind of bid is precisely the kind the economist wants to launder out as illegitimate. Likewise, the fact that the bid that is entered depends on the person's willingness-to-pay given all other financial commitments – that she is a bit stretched – is also just as it should be. The measure is of precisely willingness-to-pay at the margin for some good, given the person's current budget constraints. That this is the case raises some large distributional issues – it means that,unmodified, the poorer you are the less your preferences count. And there are standard ways in which that can be addressed by readjusting benefits and costs of to overcome that problem. I focus here however on a different problem that is being raised, namely the worry about relevance.

While all might be well from the neoclassical economist's point of view there is a procedural issue here about whether this is the kind of way in which the issue should be addressed. Should 'the future of the Levels' depend on some particular facts about individuals' lives? The problem here is in part one about the different goods involved. The neoclassical picture of the economic agent is one of a decision maker who gains welfare benefits from a set of market and non-market goods under a budgetary constraint. The individual has preferences for a variety of goods – shoes for the children, meals, a drink at the pub and so on – that have to be paid out of a 'stretched' budget. The contingent valuer is asking the person to simply consider another good – Pevensey Levels – alongside these so that the welfare benefit of this can be included along with the others – shoes, meals, drinks and the like. The response 'I couldn't afford to give *anything*' just shows that the answer is that, given existing constraints, any welfare benefit for this good is less pressing than for other goods. But that, I take it, is the 'misinterpretation' the respondents worry about.

The worry is that accidental matters about how much I might be able to afford out of my own current budget comparing the Levels to other items of daily consumption I have is not the right way of approaching the issue. Those are not the relevant facts. The Levels are a public good, currently open to being enjoyed, or, in misleading jargon, 'consumed' by everyone. The proper way to resolve the issue is to consider how much we want to resource publicly that project against other public projects, given the particular goods it involves. It is a matter of public argument about values and the relative merits of different public projects. They form the proper and relevant comparison set, not my bundle of personal goods.[14]

The problem of relevance arises here as part of a more general problem about the whole procedure of economic valuation and the process of cost benefit analysis it informs, namely that it is reason-blind.[15] It is preferences that count, not the reasons why the preferences are held. The strength and weakness of the *intensity* of a preference count, but the strength and weakness of the *reasons* for a preference do not. Preferences grounded in aesthetic, scientific and communitarian judgements about a site are treated as on par with preferences for this or that flavour of ice cream. Preferences are treated as expressions of mere taste to be priced and weighed one with the other. It offers conflict resolution and policy without rational assessment and debate. But that appears to make the process far from being rational, but, rather, whimsical. The point is made in the discussions of the respondents: '... if it goes out of fashion, it's in danger all the time, isn't it? If the price drops, nobody's going to be interested. That aspect, that way of thinking, is not really on is it? You can't put it on the Stock Market really. It's our very existence. It's our future' (ibid., p. 45).

The problem about the appeal to preferences here in terms of intensity is brought out by another question asked as a supplement to the main contingent valuation on Pevensey Levels. The respondents were asked to rank the following list of habitat types in order of preference: a) coasts, cliffs and beaches; b) woodland; c) meadows, fields and heaths; d) rivers, lakes and ponds; e) wetlands (fens, bogs and marshes) (Willis and Corkindale, Questionnaire Annex, p. 2).[16] The question is an odd one and it is not surprising that many respondents had difficulty in interpreting it.[17] The types themselves follow no clear rationale. It is not clear from the habitat point of view why cliffs and beaches should be classed together rather than cliff and heath or meadow, indeed why they should be classed with any of these. Nor is it clear that it makes sense to rank by types: given a choice of specific places individuals are likely to place some wetlands over some beaches and vice versa. However, putting these problems aside, it is not surprising that, given the question, wetlands came out as the least preferred and the coasts, cliffs and beaches the most preferred, followed by woodlands. If you asked me which I 'enjoy or like' the most I'd give the same answer: I climb on sea cliffs, play with my children on the beach and walk in woodlands. And the response to marshes – 'It's wet. It's muddy. It's horrible'(Burgess et al. 1995, p. 67) – is one with which I have some sympathy. There are specific wetlands for which I have great affection and which have a special place in my life. However, there are biologically significant wetlands I have visited which, despite the best efforts of friends to instill an appreciation, I still find just muddy, flat, and insect ridden. They are, like particular pieces of classical music, an acquired taste, one which I have never really acquired.

However, I would not want decisions to be made on the basis of these preferences. My particular preferences are not what should count here. Indeed when it comes to my own likes I can be fairly tough on them. I am, for example, unhappy with what the climbing community has done to cliffs in

Britain, both in terms of the visual impact through the practice of bolting which has despoiled some of the most beautiful limestone crags in the UK and the practice of gardening routes which has had an impact upon the fragile plant ecology of many cliffs. Correspondingly I am at least open to the suggestion that my likes should be overridden, that it might be better if there were fewer access agreements to cliffs held by wildlife trusts than there are at present, even though that would have an impact on my enjoyment, including that which I gain through climbing on bolted well-gardened crags. On the other hand, while I do not get much enjoyment from many wetlands or fully appreciate their qualities, my biologist friends tell me that they are very significant wildlife habitats. I have had conversations with artists who talk of the subtle shifting colours in the landscape and I know communities and people for whom particular marshes have a very special place in their lives and history: that will do for me. My judgement is that they should be protected. Which in particular should be given priority I am happy to trust to some degree to the advice of biologists and naturalist, professional and amateur, and to the voice of people who have practical knowledge of and special relations to different places. They have the relevant knowledge and the relevant particular relation to the place. My uninformed preference is not of much relevance.

The problem of the reason-blindness in cost-benefit analysis is particularly evident in this issue of uninformed preferences. Consider the following worries of a non-local non-visitors to the Levels had with the bidding questions:

> I said to the interviewer, perhaps I should read up on Pevensey. She said no, you don't need to do that, that you needed if you like uniformed opinion to look at in a new light ... So it depends on *your* approach. If you wanted just the public reaction then that's fair enough. But if you wanted a more informed reply, I think people have got to be put on their guard to think about it and to study what the objectives are ...

> ... without all the information you can't get the real answer ...

> It is pretty difficult to have real passionate feelings about something you know nothing about.

> It's not my area. I thought that's a question that cannot be answered. I didn't answer it. Well, I needed to know a lot more ...' (ibid., pp. 72–4).

The responses here are quite proper ones. It is not elitist or paternalist to demand informed preferences: it is what is demanded of rational procedure for dealing with the problems and it is not just 'elites' that have that knowledge. So also do local people. Information is required because it is an issue of argument about the merits of public projects, not about the aggregation of any preference, uninformed with informed.

This point about information is rehearsed in the literature. It is recognised in the literature on economic valuation that changes in the quantity and quality of information one presents, and indeed the form in which one presents it, will alter responses in willingness-to-pay surveys about environmental goods. Generally the better the information and its presentation the higher the bid. Now, one response here is to accept the point that the respondents above make – that it is informed preferences that count.[18] The issue then becomes one of how to incorporate information. A variety of methods for eliciting informed responses to contingent valuation surveys has been suggested, from attempting to inform the respondents to the use of expert groups. However, these responses are unsatisfactory for two related reasons; one practical, the other theoretical.

The practical problem is that any valuation becomes an artifact of the survey: the price you get out depends on the information and presentation you put in. This is a consequence of the theoretical point, that both price and preference are irrelevant once one moves to informed preferences. The reason why the preferences of the informed respondent count is that they are in a better position to make judgements about the value of different places and habitats. What is important is not any preference or price that is put upon the habitats, but the soundness of the information and reasons a person has for valuing the habitat. What matters in the valuations is not the preference, but the quality of reasons and information. To offer the informants information is to transform an exercise in eliciting monetary valuation into an occasion for educating the respondent. In so far as it has value it is not in virtue of its being an exercise in economic valuation but in its mimicking the proper form of decision making here, public debate.

5   Responding to the market

I want to finish by returning to the general underpinning of the neoclassical project that our environmental problems are a consequence of the existence of unpriced goods in the market. The project is related to what in the end I suspect is the strongest case for contingent valuation – the pragmatic argument. Contingent valuation serves to justify policies in economic terms; to make the sums come out right. That is after all what the Pevensey Levels survey did. The authors could conclude that the wildlife enhancement scheme was justified. Now, in a world dominated by market norms, that looks like a powerful argument. However, it is, I think, deeply mistaken. It is far from clear that in the long term coming up with unbelievable financial figures is the best way of protecting environmental goods. This is not just because the figures are unbelievable – and pragmatically contingent valuation has limited impact on those it attempts to move. The very treatment of the site in terms of commercial norms itself is part of our environmental crisis.

The issue of environmental policy is in part one of market boundaries. As I noted at the outset of the paper there is a general tendency in the modern world for the domains of commerce to expand: a series of non-market goods such as the human body, especially the womb, academic knowledge, libraries, educational and cultural goods, political deliberation and personal relationships are being subjected either to direct commodification or to the introduction of market norms. The boundaries that separate a 'free' unpriced world of knowledge, the body and so on from those of the market are being eroded. The appropriate response to the erosion of such boundaries is not to make sure that, as they disappear, the best price is achieved. It is rather to resist the disappearance of the proper boundaries between the different spheres. For example, it is neither an ethically nor pragmatically adequate response to commercial surrogacy to work out good commercial rents for wombs, rather than resist commercialisation. The same is true of environmental goods. It may be the case that the environment is unpriced and in a world in which market norms predominate this might be a problem. But strategically it is a move in the wrong direction to accept the disappearance of boundaries and simply look for a price. Protection of our environment is best served not by bringing the environment into a surrogate version of the commercial world, but by its protection as a sphere outside the world of commodity exchange and its norms. We best serve environmental goals by resisting the spread of market norms and giving them protection in terms of directly ethical, aesthetic, political and scientific standards, stated and understood in those terms. We then have a problem of working within limits we decide upon and this raises real resource issues, but these are issues for political debate and judgement, not for pricing.[19]

Notes

1   This resistance is noted by practitioners of contingent valuation: Schulz and Schulz, for example, note the significance of 'ethical reservations against the monetarization of environmental assets. Time and again, cost-benefit analysts are confronted with the question as to why phenomena such as the dying of forests and the pollution of the North and Baltic Seas, which are obviously undesirable, should require any economic-costs calculations ... [A]n intact environment is "a value in itself" and should not be tainted by an association with money' (Schultz and Schultz, 1991, p. 42).

2   For an excellent general discussion of the problem of market boundaries which touches on the environmental case see Keat, 1993; Walzer, 1983, ch. 4; Anderson, 1990.

3   I take this distinction between those two different modes of market colonisation from Keat, 1993 and 1997.

4   I criticise the first in O'Neill, 1995a.

5   Using the lower base level of £0.01 as current willingness-to-pay, the benefit of the WES comes out £111,902. The payment for the WES amounts to £147,700 for the year 1993/4. However, the authors claim that, since the study excludes those over 60 km distant, the difference might be taken into account, so one could expect the benefits of the scheme to outweigh the costs.

6   The environmental economist concerned with justifying environmentally beneficial policies and projects could note here that the critics of contingent valuation provide ammunition for players on the scene who have a very different agenda. The Treasury and Department of Environment in the UK do not employ contingent valuations of 'non-use' values in appraising projects and have expressed scepticism about their use (Treasury/Department of Environment, Volume II, paras 9 and 1938–40). Environmental economists have suggested to me that this has more to do with the fact that too many environmental projects come out as economically justifiable given contingent valuation than it does with any substantive criticism. Whether or not this is true, the environmental critic of contingent valuation does need to be aware that not all who share scepticism about the practice do so for the same reasons.

7   This is the interpretation suggested for example in McKirahan, 1994, ch. 19. This interpretation is also found in Montaigne's gloss on the Darius story in Montaigne, 1957. Criticism of the argument from variation is to be found in Aristotle *Nicomachean Ethics* I.3 and V.6. However, Alan Holland has pointed out to me that the actual context in Herodotus does not actually make this point, but rather a more simple one. The point of the Darius anecdote is to show the madness of Cambyses in his assaults on the customs of those he conquered. The story is open to an elaboration that supports Holland's 'queerness' objection to contingent valuation (Holland, 1995).

8   Cf. Raz (1986, pp. 345ff.) on the concept of what he calls 'constitutive incommensurabilities'. For a discussion of this point see O'Neill, 1993a, pp. 118–122. To the charge that this position might be essentialist my response is – perhaps it is, but there is nothing as such wrong with that: see O'Neill, 1995b, especially pp. 270–1.

9   For an elaboration of this point see O'Neill, 1993b; O'Neill and Holland, forthcoming; O'Neill, forthcoming.

10 The issue of who has property rights has long been recognised as a problem in the economic literature on environmental valuation. The distribution of property rights makes a different to agents' valuations: a standard problem in the literature is that individuals are unlikely to respond to a willingness-to-pay question, or to put in a low bid, if they believe the good in question is already theirs. The question of willingness to accept compensation becomes the relevant question. However, there is a reluctance amongst environmental economists to ask that question since the value is higher: not only are individuals less averse, but unlike the willingness-to-pay question it is unconstrained by any budget. This is a recognised problem that raises practical issues in environmental valuation in systems of property rights such as that in Scandinavian countries where there is common 'everyman's right' or free access to nature: to ask people how much they would be willing to pay for such access is to ask a hypothetical question that calls into question those common rights (see Mantymaa, Ovaskainen and Sievanen, 1992, p. 86). The issue raised by the respondents to the Pevensey Levels question raise issues of property rights that are rather the obverse of this traditional problem.

11 See Marx's comment (1972, p. 776): 'a whole society, a nation, or even all simultaneously existing societies taken together, are not the owners of the globe. They are only its possessors, its usufructuaries, and, like *bone patres familias*, they must hand it down to succeeding generations in an improved condition.'

12 For typical uses of the phrase see Pigou, 1952, p. 11 and Pearce et al., 1989, ch. 3.

13 Protest bids are treated as unjustified refusals to face up to hard evaluative choices. In defending this it is commonly noted that to put a money value on an object is not to say that money is a supreme value and apparently assumed that this same mistake is being made by the respondents (see, for example, Griffin, 1977, p. 52; Pearce and Turnerr, 1990, p. 121). This rather misses the point of those who object to treating all goods as if they could have a price. To treat price as a neutral measuring device and acts of buying and selling like exercises in the use of a tape measure is to fail to appreciate that acts of exchange are social acts with social meanings.

14 The point is implicit in the question, which asks for a willingness-to-pay 'bearing in mind that there are many worthwhile nature conservation programmes in England which you might wish to support'. Other conservation projects do form part of the right comparison set.

15 Cost benefit analysis inherits here the reason-blindness of the market itself. That the market is blind to the values and reasons that individuals offer for their preferences and choices is often celebrated by liberal

defenders of the market – it renders the market neutral between conceptions of the good. The market is an arational and amoral institution and as such allows individuals with radically different ends to cooperate in the pursuit of those ends. For a discussion see O'Neill, 1995c.

16 They were asked in addition to assess each in terms of how much they were perceived to be under threat.

17 In discussion there was some disagreement about how the question should be interpreted – as about likes or about judgements (Burgess et al., 1995, p. 40).

18 There are good theoretical reasons for this move: if a preference satisfaction account of welfare is to be used then it is much more plausible to characterise welfare as the satisfaction of informed preferences rather than preferences per se (see Griffin, 1986). My own view is that this account of welfare is still inadequate: see O'Neill, 1993a, ch. 5. Interestingly, the authors of the contingent valuation report on Pevensey Levels express some scepticism that information is relevant, although they appear to be unconscious of the possible implications that would have for their own inquiries (Willis et al., 1995, p. 85).

19 My thanks to Jackie Burgess for the material on Pevensey Levels.

References

Aldred, J. (1995), *Value Conflicts in Environmental Decision Making*, draft of PhD: Cambridge.

Anderson, E. (1990), 'The Ethical Limits of the Market', *Economics and Philosophy*, No. 6, pp. 179–206.

Arrow, K. (1984), 'Limited Knowledge and Economic Analysis' in *The Economics of Information*, Harvard University Press: Cambridge, Mass.

Burgess, J., Clark, J. and Harrison, C. (1986), *Valuing Nature: What Lies Behind Responses to Contingent Valuation Surveys?*, UCL: London.

Griffin, J. (1986), *Well-Being*, Clarendon Press: Oxford.

Griffin, J. (1977), 'Are There Incommensurable Values?', *Philosophy and Public Affairs*, pp. 739–59.

Holland, A. (1995), 'The Assumptions of Cost-Benefit Analysis: A Philosopher's View' in K. Willis and J. Corkindale (eds), *Environmental Valuation: New Perspectives*, CAB International: Wallingford.

Keat, R. (1993), 'The Moral Boundaries of the Market' in C. Crouch and D. Marquand (eds), *Ethics and Markets*, Blackwell: Oxford.

Keat, R. (1997), 'Colonisation by the Market: Walzer on Recognition' *Journal of Political Philosophy*, Vol. 5, No. 1, pp. 93–107.

McKirahan, R. (1994), *Philosophy Before Socrates*, Hackett: Indianapolis.

Mantymaa, E., Ovaskainen, V. and Sievanen, T. (1992), 'Finland' in S. Navrud (ed.), *Pricing the European Environment*, Scandinavian University Press: Oslo.

Marx, K. (1972), *Capital*, III, Lawrence and Wishart: London.

Montaigne, M. (1957), 'Of Custom and Not Easily Changing an Accepted Law' in trans. D. Frame, *The Complete Essays*, I.23, Stanford University Press: Stanford.

O'Neill, J. (1993a), *Ecology, Policy and Politics: Human Well-Being and the Natural World*, Routledge: London.

O'Neill, J. (1993b), 'Future Generations: Present Harms', *Philosophy*, 68, pp. 35–51.

O'Neill, J. (1995a), 'Public Choice, Institutional Economics, Environmental Goods', *Environmental Politics*, 4, pp. 197–218.

O'Neill, J. (1995b), 'Essences and Markets', *The Monist*, 78, pp. 258–275.

O'Neill, J. (1995c), 'Polity, Economy, Neutrality', *Political Studies*, 43, pp. 414–431.

O'Neill, J. (forthcoming), 'Time, Narrative and Environmental Politics' in R. Glottlieb (ed.), *New Perspectives in Environmental Politics*, Routledge: London.

O'Neill, J. and Holland, A. (forthcoming), 'The Ecological Integrity of Nature over Time: Some Problems', *Global Bioethics*.

Pearce et al. (1989), *Blueprint for a Green Economy*, Earthscan:London.

Pearce, D. and Turner, K. (1990), *Economics of Natural Resources*, Harvester: New York.

Pigou, A. (1952), *The Economics of Welfare*, 4th edn, Macmillan: London.

Raz, J. (1986), *The Morality of Freedom*, Clarendon: Oxford.

Schultz, W. and E. (1991), 'Germany' in J.Ph. Barde and D.W. Pearce (eds), *Valuing the Environment: Six Case Studies*, Earthscan: London.

Treasury/Department of Environment (1995), *Report from the Select Committee on Sustainable Development*, HMSO: London.

Walzer, M. (1983), *Spheres of Justice*, Martin Robertson: London.

Willis, K. and Corkindale (eds) (1995), *Environmental Valuation: New Perspectives*, CAB International: Wallingford.

Willis, K., Garrod, G. and Benson, J. (1995), *Wildlife Enhancement Scheme: a Contingent Valuation Study of Pevensey Levels*, English Nature: Peterborough.

8 Environmental goods and market boundaries: a response to O'Neill

Russell Keat

1 Introduction: economism and the environment

O'Neill wishes to reject what is, according to Arrow, the distinctively neoclassical view that environmental problems are due to the absence of prices for environmental goods and harms and hence that the solution to these problems lies either in extending property rights over unowned items, or in the use of cost-benefit analysis (CBA). The latter involves constructing surrogate prices for environmental goods and harms by methods such as contingent valuation (CV), in which people are asked willingness-to-pay (WTP) questions designed to elicit how much value they place on specific features of the environment.

O'Neill argues that this way of conceiving environmental problems and their solution is radically mistaken. Environmental problems are (to a significant extent) due to a failure to recognise and enforce legitimate market boundaries – to the application of market mechanisms and/or norms in domains where they are inappropriate, as they are in those of personal relationships such as love and friendship. To attempt to solve these problems by the extension of property rights or the use of CBA is thus to employ devices whose theoretical rationale is the source of the very problems they are intended to resolve. In particular, he argues, one should not regard the use of the latter device as a significant departure from the market-based rationale of the former. For although CBA is an alternative to market *mechanisms*, it retains the central *norms* and *meanings* of market procedures and its application to environmental decisions is thus equally to be seen as an inappropriate extension of the 'market' domain.[1]

These opposing views are reflected in contrasting attitudes towards CV respondents who display resistance towards WTP questions by making 'protest bids', or refusing to answer at all, or having serious misgivings about having provided answers. For the neoclassical environmental economist, they are the villains of the piece, whose responses must be ignored, carefully 'laundered' etc. – they are refusing or failing to play the right game. For O'Neill, by contrast, they are the heroes: their reasons for refusing the game point to its fundamentally defective character and are hence worthy of sympathetic theoretical articulation and defence. Drawing on a study of CV respondents involved in the evaluation of a wildlife enhancement scheme for the Pevensey Levels, he interprets their objections to (what I shall henceforth term) the *economistic* approach as appealing to three main considerations:

i   that features of the environment are often constitutive of – i.e., are amongst the conditions of possibility for – valued social relations, so that the loss or destruction of the former wreaks similar havoc on the latter;

ii   that to be asked or required to put a price on one's basic commitments – including those to both ethical values and people – is to be asked to betray those commitments, whose significance to one's life is marked precisely by one's refusal to put a price on them and hence to be 'bought off'; and

iii that the procedure of CBA/CV is inappropriate since it is 'reason-blind': its reliance upon 'mere preferences' ignores the need for relevant information and excludes processes of reasoning and the critical evaluation of arguments.

I agree with O'Neill that one may examine these issues about environmental decisions fruitfully in terms of market boundaries; that from this perspective, the use of CBA is to be seen as an endorsement, rather than a rejection, of the market; and that the economistic approach to the environment is indeed radically mistaken, not least because it wrongly assumes that the relevant sources or forms of human wellbeing are priceable. However, there are also certain aspects of the position for which he argues, as well as aspects of the arguments he provides for these, with which I disagree. Briefly stated, my view is that O'Neill's reasons for opposing the economistic approach prove both too *much* and too *little*. Too much, since they imply not so much a specific limitation on the market (and/or its analogues) in relation to the environment as its wholesale rejection and hence undermine the project of conceiving of these issues in terms of market *boundaries* – this objection applies primarily to (iii) above: too little, since the protection they might offer for the environment is very fragile and might equally turn out to support its degradation, due to their highly contingent and anthropocentric character – this objection applies primarily to (i) above. I shall now argue for each of these objections in turn.

2 Proving too much – breaking with the boundaries framework

To see what is wrong with O'Neill's arguments here, it may be helpful first to indicate what is distinctive about the broad framework within which these are conducted – that of market boundaries – by contrasting it with two others often deployed in exploring the problematic relations between market procedures and the environment.

a At one extreme there is the (neoclassical) *economistic* framework, according to which environmental problems are to be conceived as cases of 'market failure', where this is defined in terms of the failure of actual markets to generate the efficient, i.e., pareto-optimal results of ideal markets. The aim of procedures such as CBA is then seen as (re-) establishing, or at least getting closer to, the efficiency of ideal markets.

b At the other extreme there is the framework of *radical critique*. According to this, the market displays a number of fundamental and systematic defects which render it wholly unacceptable as a means of organising any significant area of social activity: it responds only to wants and not to needs; it transforms individually rational actions into collectively irrational outcomes; it relies upon morally undesirable forms of motivation; and so on. From this standpoint, the market's damaging effects on the environment are seen simply as exemplifying these generic defects.

c In between these two is the *market boundaries* framework, which on the one hand resists the totalising nature of the radical critique (b), yet on the other regards (a), the economistic framework, as failing adequately to characterise the nature of the market's problematic effects on the environment and hence also to provide adequate solutions for these. The environment is thus seen as one amongst a number of cases where boundaries or limits need to be placed around the operation of the market and where the need to do so is not fully captured by (a)'s limited conception of 'failure'. But the market is not wholly or inherently defective: it is acceptable when kept 'in its proper place', applied to the organisation of *certain* social activities, but not to that of *others*, and/or only when subject to various forms of regulation etc.

O'Neill represents his own arguments against economistic solutions to environmental problems as belonging to (c), the boundaries framework. But,

I shall now argue, at least one of these arguments – (iii) above, the 'reason-blindness' objection to CBA – fails to meet a crucial requirement for arguments constructed within this framework: that the reasons provided for excluding some specific item or set of items from market procedures must not be such that, when generalised, they would imply that the market is inappropriate for all (or most) kinds of item. Unless this condition is met, the boundaries framework (c) will collapse into (b), that of radical critique: an argument supposedly designed to exclude some (relatively limited) set of items will turn out to exclude so many that it will no longer be plausible to claim that 'the market is appropriate in many areas but not *this* one'.[2] To see why objecting to the 'reason-blindness' of the preferences utilised in CBA/CV fails to meet this requirement,[3] we can start by noting, as O'Neill himself does, that reason-blindness is not just a feature of CBA, but of markets quite generally: markets respond to consumers' preferences whatever these may be and based on whatever reasons. The market is a procedure in which 'no questions are asked' of the consumers whose preferences – indicated by their willingness-to-pay – determine what is to be produced, at what price and so on. There is no need for consumers to show that their preferences are well-founded, based on morally acceptable desires or empirically well-informed beliefs: no such positive judgements have to be made as a condition for these preferences' playing their part in the decision-making process of the market.

Yet if this is so, any appeal to reason-blindness *simpliciter* as a ground for rejecting economistic (i.e., market-like) procedures for making environmental decisions must imply that the market is a thoroughly inappropriate procedure for making *any* kinds of decisions, whether or not these involve environmental issues. In other words, the objection no longer belongs to the boundaries framework, but to that of radical critique. The same applies to O'Neill's later comments about what should replace the reason-blind procedure of CBA (p. 124). He claims that once we accept that 'preferences alone' are not enough, and that one needs relevant information and the critical evaluation of possible reasons etc., we will also see that neither preferences nor prices are of any relevance at all: the appropriate form of public debate will replace altogether the economistic reliance on preferences and WTP. But if this is so, the same should presumably hold in every other case in which market procedures are utilised: once their blindness to reasons has been remedied, there will be no need for preferences to be 'registered' at all – a view which quite clearly belongs to (at least some versions of) the radical critique of the market and the advocacy of its wholesale replacement by 'democratic decision-making' or the like.

To avoid this implication – and hence the collapse of (c) into (b) – one would have to show that there was something special about environmental decisions which made reason-blindness inappropriate for them but not for (at least many kinds of) other cases. That is, the objection would have to be restated so that it was no longer to reason-blindness as such, but to reason-blindness in a certain kind of case as distinct from others (i.e., from those for which the market is appropriate, or at least not inappropriate on *these* grounds). To see what might be involved in doing this, it will be helpful first to consider briefly how those who defend the market as an appropriate procedure in many kinds of case might justify its distinctively reason-blind character.

Here we need to distinguish two such justifications, only the second of which will prove relevant. The first – which might be termed 'liberal' – appeals to the value of individual autonomy and/or the principle of non-paternalism. What is admirable about the reason-blindness of market procedures is that individuals are left free to make their own judgements, without being required to justify these to anyone else: in particular, no authority is attributed to anyone but the agents themselves to make judgements about what they 'should' prefer. By contrast, the second justification – which might be termed 'welfarist' – appeals to the claim that each individual is the best judge of their own wellbeing and of what will be most likely to contribute to it. Here the overall rationale for the market is that it is a procedure which maximises aggregate individual wellbeing;[4] and it can, as it were, afford to be 'reason-blind', in the sense described above, precisely because one can rely on individuals making better judgements about their own wellbeing than anyone else can. Thus individuals' preferences are to be left unevaluated by others not because this guarantees freedom or autonomy, but because it makes it more likely that the market procedure will succeed in its aim of maximising aggregate individual wellbeing.[5]

Now, given that O'Neill's target is the neoclassical economism of CBA, with its conceptual and historical connections to the welfarist rather than the liberal justification for the market, it is the welfarist 'best judge' rationale for reason-blindness that is more relevant to explore here. Hence, to argue that reason-blindness is objectionable in environmental cases, whilst not being generally so, one would have to show that there is something about the way in which the environment contributes to human wellbeing which ('exceptionally') makes it significantly less likely that individuals will prove to be the best judges of *its* contribution to their wellbeing than they are of contributions from other sources. On the face of it, however, this is a rather implausible claim to make: it's hard to see why, in general, people should be radically less able to make competent judgements about this than about most other potential sources of their wellbeing.[6]

So it seems difficult to find a version of the reason-blindness objection to environmental CBA which both remains within the boundaries framework *and* is consistent with the relevant rationale for the market's reason-blindness in general. Or rather – and this is a crucial point – it is difficult to do so if one accepts that the only 'value' of the environment consists in its contribution to human wellbeing, i.e., if one adopts an anthropocentric form of

135

environmentalism. If, however, one rejects the anthropocentric standpoint, the difficulty I have been trying to demonstrate is far less taxing – indeed I would suggest that it is wholly removed. For then one could argue 'quite easily' that there is indeed a good reason for reason-blindness being 'peculiarly' inappropriate for environmental decisions: namely, that since in these cases we are not – or should not – be concerned exclusively with human wellbeing, the standard welfarist rationale for reason-blindness no longer applies. The 'fact' that each individual is the best judge of their own wellbeing in no way implies that they are likewise – without further debate about their (quite possibly uninformed) preferences – the 'best judge' of what is best for, or rightly done to, the environment, conceived as something whose value does not consist (exclusively) in its contribution to human wellbeing; and certainly there is no reason to expect that people's answers to questions designed to elicit their preferences with respect to their own wellbeing will generate rationally justifiable 'preferences' concerning the latter.

In other words, even if the 'best judge' assumption does provide a plausible justification for reason-blindness where human wellbeing alone is concerned, it does not thereby do so where this is not the case. Instead, it would seem entirely reasonable that individuals should, in effect, be required to justify their preferences, in at least the following respect: they should be able to show that what they prefer in terms of their own wellbeing is compatible with (what might somewhat contentiously be called) the 'wellbeing' of the environment – and if this cannot be shown, then those preferences will have to be adjusted so as to meet this additional, non-anthropocentric requirement. In this context, public debate of the kind O'Neill envisages will indeed be entirely appropriate – albeit couched in at least partly different terms from those he endorses. For O'Neill's position (and that of his favoured CBA respondents, at least as he interprets/represents them) is resolutely anthropocentric;[7] so he cannot adopt the restatement of the reason-blindness objection I have been suggesting as a way of remaining within the boundaries framework he has chosen to adopt.

I shall say more about the anthropocentrism issue later. But first I want to note that O'Neill has available to him another ground for objecting to the economism of environmental CBA, one that remains within the anthropocentric standpoint but which is not to do with reason-blindness. This is based on his rejection of a further assumption made by the neoclassicist, namely that it is possible to 'put a price' on all sources of human wellbeing and hence that it makes sense to ask the standard WTP questions employed in CV. Of course, as with the reason-blindness objection, the boundaries framework requires him to show that there is something 'special' about the environmental contributions to wellbeing which makes the 'normal' priceability condition for market procedures inoperative or inappropriate. But here he is on rather stronger ground than in his objection to reason-blindness – and indeed precisely 'because' of his anthropocentrism, or at least of the particular version of this he relies on. For O'Neill argues that the value of the environment to humans derives (at least partly) from its 'constitutive' role in certain kinds of personal and/or social relationships, whose own value cannot itself be represented by market prices and which would, indeed, be undermined or corrupted by the very attempt to do so. Hence the market-like pricing required by CBA is inappropriate here – not only for these personal/social relationships, but also for the environment, via its constitutive function for these.[8]

Unlike the reason-blindness objection, the above argument succeeds in meeting the 'non-generalisability' requirement for objections within the boundaries framework, since it can plausibly be claimed that by no means all or even most market transactions or decisions involve, at least to a significant degree, such attempts to 'put prices on unpriceable human relationships' of the kind O'Neill clearly has in mind here. Further, as I indicated earlier, I am quite happy to accept O'Neill's view that 'sources of human wellbeing' of this kind are not properly conceived as priceable. I am not, however, so happy to accept a 'defence' of the environment against market expansion which relies so heavily on these – or indeed any anthropocentric – claims.[9] So here I turn to my second overall objection: that 'too little' is proved by these.

3   Proving too little: commitments, right(s) and human wellbeing

O'Neill acknowledges that in many cases the use of CBA does lead to the 'right' decisions – i.e., environmentally-protective ones. But he thinks nonetheless that it should be rejected, not only for the reasons I have already discussed, but also because the protection offered to the environment by this means is unduly fragile, too much tied to contingencies – such as how much money environmentally-friendly respondents happen to have, which will then affect how much they are willing to pay to preserve it. Yet it seems that a similar objection applies to his own (and his favoured CV respondents') appeal to (i) above – the constitutive role of the environment in valued social relations. For this combines anthropocentrism with historical and cultural contingency in a way that likewise offers no secure protection to the environment.

Consider, for example, the claim that some particular feature of the environment should be protected because it is or has been the 'place' in which such valued social relations and the activities associated with these are or have been located (pp. 118–9). And bear in mind that, on O'Neill's account, the characteristics of such places will themselves typically have been shaped by (the consequences of) various human activities, rather than being those of 'pure' nature or wilderness. It then seems clear that protecting such places simply because they play an essential part in social relations – whose value cannot be represented in monetary terms on pain of the charge of moral betrayal – will only fortuitously protect the environment from human depredations.

For suppose – amongst many similar possibilities – that some such 'valued place' had itself been formed through a forest-clearance that destroyed the habitat of a population of some species. Then, other things being equal, to preserve that place is to maintain that species' disadvantage; and to remedy this situation, from the standpoint of the species, the 'place' would itself have to be re-fashioned, in ways that will then remove the basis for those valued social relations.[10] If it is right to protect that species, it would be wrong to preserve the conditions of possibility for those social relations; and there is clearly no guarantee that by protecting places for the latter, we will be doing the right thing for the former.[11]

The basic problem with O'Neill's appeal to social relationships is that for him the issue is about preserving the (environmental) conditions which make valued social relations possible, without any reference to whether such relations are themselves consistent with the environment's own 'conditions of possibility'. There is no reason to believe that the two sets of conditions will coincide; and when they do, this will itself be a coincidence. To avoid such thoroughgoing contingency, 'valued' social relations must be subject to certain constraints and these constraints must not be restricted to those concerning the environment's value as a contributor to human wellbeing. So although I agree with O'Neill, contra the proponents of CBA, that it is both conceptually and ethically objectionable to ask people to 'put a price' on the commitments they have to others, as expressed through the character of their social and personal relations with one another, I do not see how such commitments can serve as an adequate basis for environmental protection. The implications for the environment of such relationships and their conditions of possibility depend on the specific 'content' of the commitment involved: commitment as such is no guarantee of anything so far as the environment is concerned.[12]

This problem with O'Neill's account also suggests that his response to the King Darius story is unsatisfactory. He notes that the point of the story is meant to be that 'customs and conventions vary' – in this case, eating vs cremating one's dead parents – and is worried that this might undermine any appeal to 'what seems queer to *us*' in opposing the expansion of market norms. His solution is to say that underlying the different local customs there may well be a deeper and (more) universal human value, namely that of 'respecting those to whom one has special ties' – a commitment which people are quite rightly unwilling to abandon 'if the price is right'. But if our concern is with the environmental impact of human practices, it will often turn out that it is the specific content of the local and varying customs that matters, not the 'shared understandings' that underlie these. Even if every culture is committed to 'respecting the dead', the different customs of cremating or eating them may make all the difference to a forest.

One way in which the upshot of the objections I have been making to O'Neill's anthropocentrism might be expressed is this: that we should treat the environment in a manner analogous to that in which we should treat a being, or set of beings, which we regard as having certain *rights* – i.e., that we should accept various kinds of constraint on possible actions towards it, even though these actions could reasonably be expected to improve individual human wellbeing. I put this suggestion in a deliberately cautious way, so as to avoid any direct – and arguably problematic – attribution of rights to non-human (and in many cases non-sentient, even non-living) beings. Nor do I wish to argue that the only or best way in which what others have referred to as the 'intrinsic value' of the environment can be articulated is in terms of 'the discourse of rights'. But I think the 'rights-analogy' is potentially illuminating in at least two respects.

First – and here I return to the issues about the market boundaries framework discussed earlier – it suggests that at least one kind of boundary argument in relation to the environment might be seen as running parallel to a certain class of such arguments in relation to humans: namely, those which appeal to various human rights as the basis for justifying limitations on the scope of market processes. The most obvious example of such rights-based limitations is the prohibition on the sale or purchase of human beings themselves, i.e., on humans being attributed the status of marketable commodities;[13] and there are several other, similarly based and widely acknowledged such limitations.

Now, rights-based arguments for market boundaries or limitations differ significantly from those which focus instead on (human) wellbeing and the 'goods' which contribute to this. The latter class of arguments basically consist in trying to show that there are certain kinds of goods which the market cannot provide and which, if it is permitted to 'attempt' to do so, it will corrupt or undermine. An obvious example would be the claim that friendship – an important source of human wellbeing – cannot flourish if conducted as if it were an exchange of commodities. In other words, these are arguments designed to show that the market 'cannot deliver (these kinds of) goods', whereas the rights-based arguments aim to show that 'the delivery of (any) goods' must be subject to certain moral constraints.[14]

So the second point suggested by the 'rights-analogy' is that not only should we reject arguments for market boundaries in relation to the environment which refer exclusively to *human* wellbeing, but also that the relevant parallel with other arguments for market boundaries that *do* refer to humans is with those couched in terms of rights rather than of goods or wellbeing – with those concerned to establish proper moral constraints on the human actions that would otherwise occur in morally unconstrained market processes. However, it may well be that O'Neill would not accept the distinction I have just made between rights– and goods-based arguments, whether for market boundaries claims or more generally; and indeed, his reasons for not doing so may themselves suggest a response to my earlier objections to his anthropocentrism.

What I have in mind here is this. O'Neill claims that social and personal relationships involving commitment, loyalty, etc. to others are significant elements or sources of human wellbeing – hence people's refusal to 'put a price' on these, to be 'bribed', is to be seen not so much, or merely, as a refusal to 'betray others', but rather, or also, as an affirmation of the value of those relationships to their own lives, of the contribution they make to their own wellbeing. One might then suggest, on this basis – albeit at the risk of over-interpretation – that for O'Neill, what I have presented as 'moral constraints' on individuals' pursuit of their own wellbeing (including the constraints stemming from the recognition of rights) should instead be seen as, at least in many cases, constitutive elements of relationships such as friendship which are themselves human goods, beneficial to each party to the relationship.

In other words, on this view there is no necessary or general conflict between individual wellbeing and moral constraint since, in effect, living in 'right' relationships with others is itself a central feature of human flourishing: it is precisely through relationships in which, inter alia, the rights and wellbeing of 'the other' are properly recognised and cared for that each individual's own wellbeing is achieved. So there can be no fundamental significance to the contrast I made above between rights-based and goods-based arguments. Furthermore, it might then seem an attractive move to extend this position so as also to include within a properly conceived account of human wellbeing what might be termed 'living in a right relationship with the environment'. Then, instead of regarding the moral constraints that I have claimed should govern this relationship as detracting from the human wellbeing that would otherwise be possible, these constraints would likewise be seen as constitutive of a humanly beneficial relationship (with the environment) and of the activities this makes available. In this way, a position which remains anthropocentric can nonetheless incorporate what I have till now been claiming requires one to reject anthropocentrism.

Now, I think it is arguable that the position I have just described is not properly categorised as anthropocentric, but I shall not argue this here. Instead I shall conclude by suggesting that, if one tries to defend this analogy between 'self-beneficial-but-right relations' with other humans and with the environment, then one will have add to one's account of the latter something that seems to be absent in O'Neill's position: namely, some (environmental) analogue of the concern for the wellbeing and/or rights of the other that is included in human relationships of friendship, commitment and so on. In other words, a 'right relationship with the *environment*' must be governed by, or incorporate, some equivalent to the concern for the other *person* that is characteristic of such human relationships – a concern that will at least sometimes require the sacrifice or limitation of one's own immediate wellbeing, even if the relationship which includes such limitations is itself of great or overriding value to one. It follows from this that the value attributed to the environment cannot consist wholly, as O'Neill implies, in the part it plays in *our* (i.e., human-with-human) relationships, since this would be the moral equivalent of those relationships themselves being thoroughly one-sided and hence not right.

Notes

1   I shall follow O'Neill in focusing on the use of CBA and especially of CV. It is arguable, in any case, that the most plausible rationale for the former solution, the extension of property rights, lies outwith the framework of neoclassical economic theory.
2   Admittedly, this way of drawing the distinction between (b) and (c) makes the difference between them partly a matter of degree; but this will not be problematic in this particular case, since the 'reason-blindness' objection has a thoroughly general character, i.e., it would imply that there are *no* items for which the market is suitable, not merely 'rather few'.
3   Notice that the same may well be true of what is implied by the Pevensey respondents' objection (i), since the dynamism of the market is 'corrosive' of many social relationships and would thus be ruled out altogether by any wholesale 'conservative' commitment to the preservation of such relationships and/or their environmental requirements.
4   Or is pareto-optimal: the differences between these two formulations do not affect the argument here.
5   In light of this welfarist rationale for reason-blindness, it is misleading for O'Neill to imply, as he sometimes does, that economism involves *no* 'commitment to values', whereas its opponents (including the Pevensey respondents) *do* have such commitments. For example, he accuses the 'economists' of asking people to betray their commitments to values by requiring a price to be put on the environment and says that disputes about the environment are not simply about resources but about values and valued entities. But the welfarist can surely reply that s/he is *also* concerned with values – i.e., the value of environmental resources for individual wellbeing, etc. Likewise, in his discussion of 'procedural' problems, O'Neill contrasts public debate about values with economistic calculations based on people's 'bundle of personal goods'. But this contrast is misleading. The welfarist is committed to a value-theory

141

according to which what matters is precisely these 'bundles of personal goods': markets and/or CBA are then proposed as devices for making decisions which will realise this value. But this is perfectly compatible making 'reasoned debate' the basis for accepting this value in the first place – it is not a view which rules out the relevance of any such debate. We may well demand reasons for accepting such values. But this does not entail that the means by which these values are realised in practice should consist in further such debate. For it might turn out instead that they can best be realised by 'reason-blind' procedures such as the market or CBA (see Keat, 1994 for a development of this argument against a similar objection to economism by Sagoff, 1988; also Keat, 1997a).

6   It might be objected that I am ignoring here the problem of future generations, which is 'peculiarly' relevant to environmental decisions. But the proponents of CBA can and often do at least attempt to include some reference to these; and it is anyway not a problem encountered only by economistic approaches.

7   Or at least, he gives no indication of wishing to rely on anything but anthropocentrically based objections to CBA; nor does he do so elsewhere, e.g., in O'Neill, 1993.

8   It might be argued, against O'Neill, that it does not follow from the fact that something (in this case the environment) contributes to non-priceable wellbeing that this contributor is itself unpriceable. Whether this objection is met by the claim of the environment's 'constitutive' role – and indeed whether this claim is itself correct – I shall leave aside here.

9   To reject anthropocentrism does not, of course, imply that human wellbeing is of no significance in valuing the environment: it implies only that the value of the environment is not wholly derivative from this.

10 But notice that 'returning nature to its pristine condition' is not the only way in which such re-fashioning could be achieved: what I am arguing for does not entail a 'wilderness ethic'.

11 A similar point applies to the Pevensey respondents' concern to 'pass on a decent heritage to their children': for all one knows, they will want to pass on the Disney theme park where they spent their honeymoon...

12 Cf. the potential ambiguity on p. 117 where O'Neill says that '*some such relationships of commitment ... are components of ... a good life*' (my italics). Does 'some' here mean 'any'? If so, caveat environment, since the commitments might not be environmentally-protective. Or does it mean 'some specific'? If so, the issue is what these should be, so as to be consistent with the protection of the environment.

13 See, e.g., the discussion of 'blocked exchanges' in ch. 4 of Walzer, 1993; I have discussed elsewhere Walzer's approach to market boundaries (1993 and 1997b).

14 On why exchange-relations undermine friendship, see Anderson, 1990; for a discussion of goods-based arguments for market boundaries, see Keat, 1996.

References

Anderson, E. (1990), 'The Ethical Limitations of the Market', *Economics and Philosophy*, No. 6, pp. 179–205.

Keat, R. (1993), 'The Moral Boundaries of the Market' in C. Crouch and D. Marquand (eds), *Ethics and the Market*, Blackwell: Oxford.

Keat, R. (1994), 'Citizens, Consumers and the Environment', *Environmental Values*, No. 3, pp. 333–50.

Keat, R. (1996), 'Delivering the Goods: Socialism, Liberalism and the Market', *Waverley Paper*, Department of Politics, University of Edinburgh.

Keat, R. (1997a), 'Values, Preferences and Neo-Classical Environmental Economics' in J. Foster (ed.), *Valuing Nature*, Routledge: London.

Keat, R. (1997b), 'Colonisation by the Market: Walzer on Recognition', *Journal of Political Philosophy*, Vol. 5, No. 1, pp. 93–107.

O'Neill, J. (1993), *Ecology, Politics and Policy*, Routledge: London.

Sagoff, M. (1988), *The Economy of the Earth*, Cambridge University Press: Cambridge.

Walzer, M. (1993), *Spheres of Justice*, Martin Robertson: London.

9 Private rights, public interests and the environment

Donald McGillivray and John Wightman

Can private rights have any significant role to play in relation to environmental protection? Traditionally, private law (mainly land law and tort law) provided the owner of land with rights to use the land and rights to prevent interference with it. But the growth of public regulation of land use during the twentieth century has seen the limitation of rights to use and the activities of pollution control and planning authorities have marginalised tort claims against interference. The conventional wisdom has been that only through the attenuation of private rights by regulation in the public interest will the environment be protected.

This view of the relative roles of private rights and public regulation has increasingly been challenged, especially by the work of economists, who, fortified by theories of regulatory failure, have argued for a larger role for private rights and the market in shaping environmental outcomes. Much of the earlier work, following Coase (1960), advocated allowing the trading of property rights in the market effectively to define environmental goals (see, e.g., Beckerman, 1990). More recent work, particularly that of Pearce et al. (1989), has emphasised the role of economic instruments (including tort liability) to attain politically defined goals (see also Jacobs, 1991). But in both cases, private law is seen as wedded to the market, whether it be by defining bundles of rights for exchange, mimicking the market in allocating liability where no exchange is possible, internalising externalities or transmitting incentives for pollution abatement through civil liability.[1]

Our aim in this paper is to explore whether a course can be steered between these two positions. We argue that problems with traditional regulation do leave space for an important role for private law, but one that is based not on its being a carrier of market goals or mechanisms, but on its providing a means of privately initiated (i.e., unofficial) challenges to official definitions of the public interest. The kind of situation we will be particularly concerned with is where a private action is brought in respect of an interference which has been licensed in some way by a public body; this is well illustrated by the recent case *Hunter v Canary Wharf* ([1996] 1 All ER 482). Residents in East London sued for damages in private nuisance respect for a variety of interferences caused by the massive docklands development. These included interference with television reception caused by the Canary Wharf tower and deposits of dust. The developers had obtained the appropriate environmental licences and planning permissions and so the issue for the court was whether private law rights could prevail over a decision taken by officials acting on behalf of the state.[2]

We argue that in situations such as these the private law action contains the potential to prise open issues about the desirability of the development and its impact on a locality, thereby increasing the accountability of and participation in the overall decision making process. Clearly, such a role goes some way beyond that usually claimed for private law. Our present concern is to assemble an argument in defence of such a role, in the hope of stimulating wider discussion of the limits of private rights in relation to the environment.

We begin by considering whether private law has the potential to express conceptions of the public interest. We then explore the various judicial approaches to licences and planning permissions in private law actions, looking especially at the role of the locality doctrine in the tort of private nuisance. We argue that this doctrine has long been used (albeit indirectly) as a means of reaching decisions on issues of public interest in private actions and that it is the spatial variation which this doctrine injects into the law which creates space for nuisance law to adopt interpretations of the public interest which may depart from those of regulators and planners. We conclude by considering the constitutional basis for private law's being used to second guess decisions taken by public authorities and argue in favour of a pluralist approach which preserves a role for private law as a means of mounting privately initiated challenges to official definitions of the public interest.

Private rights and public interests

Most private law is private in the sense that it protects the economic or bodily interests of litigants. Contract rights are enforceable by a party to the contract, and tort claims arise where a private interest has been infringed, typically resulting in compensation for personal injury or damage to property. Normally, such awards benefit no one but the plaintiff. In relation to the torts concerning interests in land, it is possible to identify a 'pure' form of privateness which is most closely approached in the tort of private nuisance. Three aspects of privateness can be distinguished. First, the nexus with land: it is interference not with any activity, but the use and enjoyment of *land* which is actionable, and traditionally an interest in land – effectively some form of private ownership – has been required. Second, it is sufficient if a *single* proprietor is affected: there is no requirement of more widespread impact and impact on others is irrelevant to whether the interference is actionable by the plaintiff. The third aspect is the requirement that there is an interference with use and enjoyment *of the plaintiff*: this includes activities affecting amenity, but not those which merely disturb or destroy the natural world without affecting a person's use.

Even where the private right is drawn in these terms, it may have the effect of protecting interests wider than the immediate concerns of the right holder. An injunction to restrain the emission of smoke or smell will normally benefit others in the neighbourhood as well as the plaintiff. However, the reach of the effect of an injunction against some polluting activity is not limited to the more obviously private interests of neighbours. There may be other classes of 'beneficiaries' who are benefited by the nuisance action. Thus, anglers may benefit through the improvement in habitat for fish from a riparian owner's success in obtaining an injunction to prevent pollution of a river by a sewage outfall, while environmentalists may approve of the outcome simply because it conserves a part of the natural world, rather than on account of any amenity use they make of the river.

These examples show how law which is purely private in form may be used to pursue wider interests. More radically, it is possible for one or more of the three aspects of privateness to be relaxed as formal components of a private law action. In this respect there are clear signs of some dilution of the traditional approach to the nexus with land in private nuisance, where occupation rather than a legally recognised interest in land is now the test (*Khorasandjian v Bush* [1993] 3 All ER 669; *Hunter v Canary Wharf* [1996] 1 All ER 482).

An important instance where a nexus with land is not essential is the protection of interests in a fishery. In the leading case *Pride of Derby and Derbyshire Angling Association v British Celanese* ([1952] ch. 149) the plaintiff angling club obtained an injunction restraining pollution of the River Derwent.[3] The club's standing to bring the action stemmed from its ownership of the fishing rights. In legal terms these rights are known as a profit a prendre, a legally-recognised species of real property right. In origin the right is similar to a right of common, protecting the right of the holder to take from an otherwise unowned source. Unlike some such rights it can exist 'in gross', that is without being attached to any neighbouring land. Originally, it was a means of preventing the unauthorised taking of fish, but is used now by angling clubs to preserve fisheries in the face of pollution of watercourses. Now that anglers normally return fish to the river, the concern is with preservation of habitat rather than the taking which was originally protected by the right. The potential to protect habitat means that fishery rights can be of use not just to those who wish to see a sustainable use made of the fish stocks but also those whose principal concern is preservation of watercourses in an unpolluted state.

Public nuisance, a curious hybrid with both civil and criminal aspects, also relaxes some of the components of the pure private right. As in the case of other continuing criminal acts, it is possible for the Attorney-General in a relator action to seek an injunction at the request of affected citizens; alternatively, the local authority can bring an action under s.222 Local Government Act 1972. A person suffering special damage, i.e., more than others who are affected, can claim damages. It therefore differs from private nuisance in two relevant respects: there is no nexus with land; and it is the general impact on a group, not on an individual, that matters, except in the case of a claim for special damage.

Public nuisance does not eclipse private nuisance as far as injunctions are concerned, because citizens have no right under public nuisance to seek the injunction: the decision lies in the hands of the Attorney-General or local authority. This means that the nexus with land remains a prerequisite for a citizen to seek an injunction for any kind of nuisance. But should this particular dimension of privateness be a sticking point? The slight loosening in *Hunter* (op. cit., p. 498) from interest in the land to 'occupation of property as a home' could go further: say, noise or pollution affecting a person's place of work or place of recreation. Of course, there will typically be an owner of land (e.g., a sports club) in whose name an action under traditional principles can be brought. But the ownership by a club here is in one sense a technicality, for it simply represents the aggregate of casual users for sporting purposes, none of whom are in occupation in the sense envisaged in *Hunter*. And although the form of ownership is lacking, the same substantive reasons for protection encompass an amenity enjoyed without being in any particular place, such as bird watching affected by deterioration in water quality or quantity essential for bird life. Here there is no citizen's remedy at present, but further development in private or public nuisance could enable an injunction to be sought in these circumstances.

Although dispensing with the nexus with land may seem unlikely, it is in fact precisely what was contained in the proposed EC directive on waste (COM(89) 282 final, modified by Amended Proposal COM(91) 219 final). The proposed waste directive applied mainly to solid wastes and sludge deposited on land. It provided for strict liability for the producers and eliminators of waste, resembling traditional private law by giving rights of action to those suffering personal injury or property damage. But it also went beyond traditional private law in two interesting ways. First, liability attached to any 'impairment' of the environment, including anything not in private ownership (important instances are air and some water-bearing strata). Second, the draft directive provided that common interest groups could bring actions, either to prevent impairment or for the cost of remedying damage. This proposal is striking, for not only is the nexus with land broken, giving common interest groups standing to apply for injunctions, but there is no need to show any interference with any human activity: impairment is sufficient.

In this respect the draft directive goes well beyond the provisions relating to statutory nuisance contained in Part III of the Environmental Protection Act 1990. These dispense with the nexus with land (action must be taken by local authorities, either on their own initiation or that of local residents), while persons 'aggrieved' may insist that a magistrate's court serve an abatement notice in respect of existing nuisances, allowing private initiation of actions. But statutory nuisance is, at least presently, concerned with matters 'prejudicial to health or a nuisance', with 'nuisance' generally understood as being defined as either a public or private nuisance at common law (*National Coal Board v Thorne* [1976] 1 WLR 543).

Of all the examples discussed, therefore, the proposed waste directive is the one furthest removed from the pure form of protecting private interests.[4] Private law can therefore be seen to encompass a spectrum of types of claim, running from the protection of orthodox individual proprietary interests to the protection of more generalised public interests in the environment, including both general amenity uses (such as bird watching or hiking) and impact on the environment which does not affect human use. If private law were extended to recognise such claims, citizens – typically acting through common interest groups – would have a legally recognised interest as custodians, rather than as owners or even users.

Specific developments in property law suggest similar possibilities of moving away from 'pure' forms of private rights. It has been argued that there is a discernable trend towards conceiving of 'a significant "equitable property" in the quality and conservation of the natural environment' (Gray, 1994, p. 188). Gray suggests the appearance of new, collective private property rights in respect of at least those parts of the natural environment considered 'ecologically imperative' (Karp, 1992–93, pp. 745ff.), making them simply non-excludable in the traditional sense. He relies heavily on examples from wilderness conservation, especially from the US, where greater political and legal salience attaches to the spiritual dimension associated with exposure to wild country (see, e.g., Nash, 1982). This dimension may be less important in the UK for various reasons, historical and geographical, but the basic point remains: the more that collective rights in property are recognised, the greater the law seems to be prepared to dilute the individualistic conception of property which has, so clearly, gained pre-eminence with the rise of the market.[5]

Private rights and public regulation

Even if it is possible to carve out a role in theory for private rights to be used to protect public or collective interests, what happens when private law clashes with regulation? The *Hunter* case raises the question of what effect the regulatory decision has on the private law claim, although the courts have yet to supply a clear answer. It is useful to explore the judicial response by distinguishing two approaches: the single-track, which seeks to align private law with regulation; and the twin-track, which permits some degree of variation between them.

A single-track approach seeks single 'right answers', so that both public and private law give the same response. This can arise in two ways. Either the private law claim yields to the public law permission, or (less commonly) the public law permission yields to the private right. The main instance of the former is the defence of statutory authority in nuisance, which effectively deprives neighbours of any claim in relation to anything done in the exercise of a statutory power as long as it is not done negligently (*Allen v Gulf Oil* [1981] AC 1001). Another example is *Budden and Albery v BP Oil* ([1980] JPL 586) where a claim in negligence against two oil companies based on their use of lead in petrol was dismissed. For the court, allowing the action would mean that valid regulations prescribing the lead content in petrol would effectively be replaced by a lower, judicially-determined standard.[6]

The other kind of single-track approach, where the public law is aligned with the private right, is much more rarely found, but can be detected in remarks in *Wheeler v Saunders* ([1995] 2 All ER 697). There, the plaintiff was successful in a private nuisance action against a neighbouring pig farmer, notwithstanding that the pig units from which the offending smell was coming had been constructed with planning permission. In the Court of Appeal, Sir John May (ibid., but see *R v Exeter City Council ex p Thomas* [1990] All ER 413) thought that:

> [If] a planning authority were with notice to grant a planning permission the inevitable consequence of which would be the creation of a nuisance, then it is well arguable that grant would be subject to judicial review on the ground of irrationality.

On this latter view, the content of the private law of nuisance is treated as circumscribing the powers of planning authorities in granting planning permissions.

The twin-track approach, in contrast, permits public and private law to give different answers to the lawfulness of an activity. This approach was adopted (with some ambiguity) in *Gillingham BC v Medway (Chatham) Dock Co* ([1992] 3 All ER 923) and (more clearly) by the majority view in *Wheeler*. In

Gillingham, the effect of a planning permission granted to a port operator to develop the port was greatly to increase the inconvenience (especially noise) caused to local residents from the increased heavy goods traffic generated by the development. Although Buckley J rejected a direct analogy between statutory authority and planning permission, he held that the permission altered the nature of the locality, so that no actionable nuisance was found. In *Wheeler* the court found that the permission did not change the nature of the locality and in *Hunter* it was held that the planning permission for the Canary Wharf tower did not authorise interference with television reception in its 'shadow'. The twin-track approach thus preserves a space for private law actions even where an interference has been licensed. So, where private law is being used (as suggested in the previous section) to pursue a public interest, there is a possibility of using it to argue for a different meaning of the public interest from that produced by the official regulatory authority. Especially in those 'twin-track' cases involving the interplay of nuisance law and planning permissions, the so-called 'locality rule' is clearly central, but on exposure to closer scrutiny the nebulous nature of this rule is exposed.

The locality rule and normative decision making

Ascribing such a clearly policy-based role to private nuisance will seem to some to reach far beyond its traditional scope. However, it is not new for private nuisance to be used in this role and it is arguable that the judges have routinely played a policy role in nuisance cases since the emergence of the locality rule in the middle of the nineteenth century.

The origin of the locality rule is usually traced to the decision of the House of Lords in 1865 in *St Helens v Tipping* ((1865) 11 HLC 642) (Brenner, 1974; Maclaren, 1983; Simpson, 1995). In the early 1860s, in two cases about brick burning the issue had arisen of whether it should be a defence in a nuisance action that the interference was for the public benefit. In *Hole v Barlow* ((1858) 140 ER 113) the court held it to be a defence, while in *Bamford v Turnley* ((1862) 3 B & S 66 per Bramwell, B., p. 84) it held it was not. *Bamford* held out the prospect that private rights would invariably prevail over public interests, but in *St Helens* the House of Lords dealt with the problem indirectly through the application of the locality rule: it was not for the judge to determine when the public interest overrode the plaintiff's rights, but an interference was less likely to amount to a nuisance in an area already subject to similar interferences.

In principle, there are two ways the court can arrive at the nature of a locality. It can be prescribed, by a regulator or judge,[7] or it can be treated merely as an empirical question, a fact about a neighbourhood which the court simply has to discover. Judges have traditionally steered clear of explicitly adopting the former view: it was and is not their job to instigate prescriptive zoning of different uses. But while courts have tended to speak of the locality as a fact, it has provided the cover – at least in marginal cases – for a more prescriptive approach, which entails the court's deciding the normative question of which uses should prevail, not just discovering what the nature of the locality is. In this respect the locality rule resembles the reasonable person test for breach of duty in negligence, which presents as an empirical enquiry (what does the reasonable person do?) what is in reality a question of what the standard of care *ought* to be.

Arguably, therefore, the locality test provides room for the court to reflect public interest considerations, but the pragmatic operation of the locality rule does not need this to be explicitly acknowledged. But the limitations of the locality rule, if conceived as a factual test, are forced into the open by the attempt to handle planning permissions; this becomes apparent if we attempt to elucidate the courts' approach to a planning permission in nuisance cases.

The judges in *Wheeler* and *Hunter* thought that planning permission could be relevant to whether an actionable nuisance existed – not on an analogy with statutory authority, but by changing the nature of the locality and thus denying what would otherwise have been actionable because of the changed locality. But how, precisely, does the planning permission bring this about? There seem to be three possibilities.

The first is that the permission will change the nature of the locality as soon as it is granted. This seems implausible: there are many permissions granted which are never acted on and valid permissions may exist for more than one development or change of use at any one time in relation to the same parcel of land. To hold that the permission operated in this way really would make it equivalent in effect to statutory power, since any preexisting rights protected by a nuisance claim would be extinguished merely by the legal act of the granting of the permission.

The second possibility is that the locality is changed when the permitted development takes place. This would turn the issue into a question of fact about the locality: what is its nature? The problem here is that in the well known cases where the locality rule has been applied the nature of the locality has been assessed *apart* from the defendant's interference. The defendant's interference itself cannot redefine the nature of the locality so as to render the interference unactionable.[8] Thus although the actions of a single defendant, such as increasing the lorry movements in *Gillingham*, may factually change the character of a neighbourhood, that change is not traditionally relevant to the locality test for the purposes of private nuisance. It would of course be possible to relax the traditional approach in the case of planning permissions, so that the interference itself could define the nature of the locality for the purposes of the locality rule. But that would be *effectively* to introduce the statutory authority principle by the back door: a special rule would be applied

151

to planning permission which would legitimate interferences from development as soon as it commenced.

The third possibility is that a planning permission changes the nature of the locality only when the factual change brought about by the permitted development has lasted long enough for potential plaintiffs to lose their rights (by prescription) to sue in nuisance; this period is usually said to be 20 years.[9] But if this is the point at which the effect of the planning permission is felt, the effect is remarkably weak. Planning permission is then no more than the start of a process of legitimating a development, the most important part of which is the existence of the interference caused by the development without challenge for 20 years. On this view planning permissions have no extra ability to override private rights since those rights are lost only in accordance with general rules about the acquisition of prescriptive rights which apply whether or not planning permission played any part in the genesis of the interference.

The upshot of the above is that as long as locality is treated as a matter of fact, the locality rule is not a coherent way of reflecting any special significance attaching to planning permissions short of the established approach to statutory authority. There does not seem to be any half way house: permissions either intrinsically legitimate an interference, or have no special weight at all in the decision on whether the interference is an actionable nuisance.[10]

The only means of finding a middle way which permits the court to give some weight to planning permission without treating it as equivalent to statutory authority is by acknowledging that judicial operation of the locality rule is in part normative – in other words, that the judge's decision *does* take a view about where the public interest lies. This way of looking at it reshapes the issue which planning permission poses for nuisance law. It is not about whether private rights (protected by nuisance law) should always be overridden by the public interest; nor about whether the judges in tort cases should be involved in prescriptive zoning at all: it is about what weight judges should give to planning permission in a balancing process which is *already* normative in character.

Seen in this way, the planning permission does not clash with some predefined right of the plaintiff. Private law in this context – private nuisance – does not typically create hard edged rights which can contradict administrative decisions. While it is usual to speak of private law as conferring private *rights*, this terminology can be misleading in the context of the land torts, which are not universally characterised by a degree of formal rationality where rules are applied relatively determinately. Though trespass to land is most like this model,[11] private nuisance is much less formally rational in character: what will amount to an actionable nuisance is difficult to predict precisely because of the policy element just discussed. In reality, when the court decides the case, it is not so much protecting a preexisting right as defining what rights the plaintiff has. In other words, 'right' describes the end result of the decision making process rather than its motivating reason.

The scope which private nuisance offers for normative or policy based decisions distinguishes it from other torts. In the tort of negligence, policy does not need to be made in the routine case, where the court typically applies the judicially-constructed standard of the reasonable man to decide whether the defendant's conduct was negligent. In nuisance, however, normative decisions are more common. There is no *general* test – equivalent to the reasonable person test in negligence – to apply, the reason being that nuisance law contains a spatial variable: what amounts to an unreasonable interference with the use and enjoyment of land will vary from place to place, principally because of the locality rule.[12] In the classic statement by Thesiger LJ in *Sturges v Bridgeman* (op. cit., p. 865), '[W]hat will be a nuisance in Belgrave Square would not necessarily be so in Bermondsey'.

This same spatial variability can also be seen in much environmental regulation. A planning permission is specific to a particular parcel of land and reflects the unique character of its spatial location. The conditions attached to discharge consents reflect the condition of the receiving watercourse at the point where the discharge enters. This contrasts with regulation outside the environmental field, which tends to have much less spatial variation. The regulation of consumer protection, or of health and safety at work, are general in impact, in that the same standards are in force nationally and (increasingly) across the European Community.

The variation along the spatial dimension of both environmental regulation and nuisance tends to weaken the case for a single track approach in which private and public law give the same answer to the question of the legality of an activity. The single-track approach seems most persuasive where there is a direct clash between some general public law provision and a general rule of private law. For example, in *Budden and Albery* the court rejected the plaintiff's argument that the defendants were liable in negligence in respect of lead in petrol notwithstanding the fact that they had complied with the regulation specifying the maximum level of lead; it would undermine the sovereignty of Parliament and result in a 'constitutional anomaly' if the claim were allowed to succeed (*Budden and Albery v BP Oil*, op. cit., per Megaw LJ, p. 587). But where both the regulatory decision and the decision in a nuisance case are tailored to local factors, the constitutional stakes are not so high; the private law decision is not in conflict with primary or secondary legislation.

Spatial variation may also help to explain the clear reluctance of judges to treat planning permissions as equivalent to statutory authority. The legitimacy of general legislative rules is clearly grounded in a representative democratic process. But spatially contingent planning permissions have a weaker democratic basis, being neither the expression of a general rule having legislative authority, nor the product of a fully democratic, deliberative procedure at local authority level.

It may seem strange to cite a problem about the democratic legitimacy of planning permissions in aid of an argument for a role for private law in shaping the public interest. Whatever the vagaries of planning permissions, surely the processes of central and local government have more claim to democratic legitimacy than can be mustered by the law of tort? We now turn to consider how far it can be justified for the legal system to embrace conflicting understandings of the public interest.

Privately initiated legal action, public interests and pluralism

How can we begin to justify the use of private law to argue for a definition of the public interest which is at odds with that reached by a public authority? Our case for a role for private law in shaping conceptions of the public interest rests on its potential to counteract regulatory failure by providing an institutional means of opening the substance of regulatory decisions to scrutiny. The model of environmental decision making we envisage is pluralist in the sense that the official decision sanctioned by public law would not necessarily be the final word on the legality of an activity. Individuals and groups could privately initiate legal proceedings in which it would be possible to challenge not only the infringement of private interests, but also the regulator's view of where the public interest lay.[13]

In a world of perfect regulation, such institutionalised second guessing would be indefensibly irrational. But given the various modes of regulatory failure – which apply to the identification of public as much as the treatment of private interests – an institutionalised pluralism is, we argue, defensible.[14] Thus it is possible to regard the whole legal response to environmental issues as comprising not just the regulatory view, buttressed by its official status, but also the outrider of unofficial legal action. In this pluralist view, private law takes its place alongside other legal mechanisms, notably private prosecutions and applications for judicial review, which are unofficial since the initiation of legal proceedings is not controlled by regulatory agencies or other parts of the state. This way of seeing unofficial legal action places emphasis not just on the outcome of litigation, but on the more indirect effects of legal action in raising the visibility of an issue, in turn leading to more rigorous scrutiny. The instigation of private action can therefore foster a wider and more informed debate about a proposed or existing development than may typically be found within routine planning procedures.

The situation in *Gillingham* provides an example of how private law could be used to challenge an official view of where the public interest lies. By 1990, the local authority was clearly less concerned about promoting employment than it had been when, in 1982, it had granted permission to redevelop the former naval dockyard. Although planning authorities can revoke permissions (if compensation is paid), in *Gillingham* the local authority was pursuing the action in public nuisance in the hope of stopping the interference from the lorries by injunction without having to pay compensation. Although this use of public nuisance was seen by the court almost as an act of bad faith, it is also possible to view it as a legitimate attempt to establish that the balance of argument about whether the harm inflicted was justified on public interest grounds had changed.[15]

The private action here can be seen as a means of revising a regulatory decision, but does not second guess the original decision at the time it was made. Rather, it provides a means of revising the balance of factors in the light of changed circumstances. This approach to the effect of the planning permission concedes that it may in some circumstances prevail over a private right. But instead of regarding it as a permanent removal of a plaintiff's rights, it sees the planning permission only as a conditional borrowing of those rights, to be returned if the calculus of public interest changes to remove the justification for the borrowing.

A further argument for a twin-track approach which permits private law's engaging with a conception of the public interest at odds with regulation draws on the justification sometimes put forward for the restrictive interpretations for the powers of public bodies. There is a tradition in administrative law of construing restrictively legislative provisions which empower officials to interfere with property rights of individuals.[16] On this view, private rights have a special status which require clear legal provisions to override them. It is possible to see the approach in *Wheeler* and *Hunter* in these terms, as instances of the court's confining the erosion of individual property rights to cases where the legal authority is wholly explicit and affected individuals are heard. Thus, in *Wheeler,* the comparison was drawn by Peter Gibson LJ between the Private Bill procedure before the enactment of a statutory power (which may be a defence to an action in nuisance), which gives affected parties rights of representation and the system for determining planning applications. The latter, he noted, afforded only limited opportunities for affected parties to obtain a remedy.[17]

It could be said that this argument, because it attaches a special significance to private rights, would not apply to attempts to use private law to protect more public interests. It is, however, possible to justify arguing that it be extended in this way. According special weight to private property rights can be seen as just one way of ascribing significance to the status quo which exists before any development or other change takes place. If our argument is accepted that private law could encompass not only amenities like bird watching (which are detached from occupation of any particular parcel of land), but also impairments of the environment which do not affect human use, then the protected legal status quo would be sufficiently broad to stretch beyond pure private interests.

The ascribing of special weight to 'public' status quo rights then rests on the claim that legal rights which people hold in common (not in the narrow legal sense) are no less deserving of such weight than rights held by individual owners. Put this way, the argument that private law claims should not be eclipsed by regulatory decisions rests not on their being purely private in nature, but on their being protective of an existing local distribution of land use. This may reflect a host of different uses and be the repository of more diffuse meanings and enduring personal attachments, all of which arguably have some claim to special consideration when the scope or legality of administrative fiat authorising their destruction is subject to legal challenge.[18]

Conclusion

In this paper we have explored a possible role for private law in relation to the environment – one which sees it as neither inevitably dominated by regulation, nor merely as the means of implementing a market based approach to environmental decision making. We have emphasised the *unofficial* nature of private law, for it is this which provides the space for common interest groups, as well as individuals, to argue for a different interpretation of the public interest from that reached by an official body.

This emphasis on the role of unofficial legal action in shaping environmental decision making is in tune with recent trends. Increasingly, the decisions of regulatory authorities and large utilities are questioned indirectly through unofficial legal action, notably judicial review. And increasing use of tort claims is also detectable, especially in relation to water pollution where information is more readily available and causation often less difficult to show than with, for example, air pollution.[19] These trends reflect not only a spreading awareness of the limitations of traditional 'command and control' regulation, but also, perhaps, a scepticism among environmental groups about the hubristic claims sometimes made for economic instruments. The kind of pluralism we advocate sees these trends not as the reflection of some dysfunctional failure in the legal system, but as a desirable development which enhances the accountability and quality of, and participation in, environmental decision making.[20]

Notes

1 Greve (1996) suggests that, in certain respects, the US courts have in fact begun to move away from a regulatory paradigm towards a

reassertion of legal rights based on common-law harms rather than environmental values. His focus, however, is limited to regulatory takings, standing rules and judicial review and does not, for example, encompass developments in property law generally. A less developed defence of property rights for environmental protection is given by Brubaker (1995).

2   The London docklands area was designated an Urban Development Area by Order (SI 1981/936) made under s.134(1) Local Government Planning and Land Act 1980, which explicitly referred to such Orders being made where 'expedient in the national interest to do so'.

3   The riparian owner was also a plaintiff in this action

4   Note that the draft directive has effectively been superseded by the Commission's Green Paper *Remedying Environmental Damage*, COM(93) 47 See also the Convention on Civil Liability for Damage Resulting from Activities Dangerous to the Environment, Lugano, 21 June, 1993, 32 ILM 1230 (1993), Art. 2, para. 7. The European Commission appears keen to keep the possibility of such legislation on the political agenda (see ENDS Report 260, September 1996, pp. 38–40).

5   A consequence of this for environmental protection, of course, might be that the successful use of trespass in cases such as *League Against Cruel Sports v Scott* [1985] 2 All ER 489 may be frustrated in the interests of a more communitarian approach to fundamental property interests.

6   Another good illustration of legislative standards effectively setting common law standards is *Cambridge Water Co v Eastern Counties Leather* [1994] 1 All ER 53, where harm was defined in accordance with the terms of the 1980 EC Drinking Water Directive.

7   It was normal in civil actions in the nineteenth century for juries to be involved where actions for damages were heard in the common law courts, but not where the action was for an injunction in Chancery. From 1875 (after the Judicature Acts) all nuisance cases where both damages and an injunction were claimed were heard in the Chancery Division and thus in the absence of juries. It is likely, therefore, that juries were only involved in determining actions relating to the locality rule during the period 1865–75 (the situation would have been different if the activity had ceased and damages only were being sought).

8   If it could, then the largest interferences – those that changed the locality – would never count as actionable nuisances

9   Since it is no defence to allege that the plaintiff 'came to the nuisance' (*Sturges v Bridgman* (1879) 11 Ch D 852) it is possible, for example, for someone living on a new housing development to bring a successful action in nuisance even where the activity complained of has been conducted for 20 years, as long as it did not interfere with the previous use (see, e.g., *Miller v Jackson* [1977] QB 966).

10 The corollary of this is that the nature of a locality will only be regarded as changed in law when potential plaintiffs have lost their rights to sue in nuisance. Put another way, if the nature of the locality really is treated purely as a matter of fact to be discovered, the locality doctrine contributes no more than the rules on the acquisition of prescriptive rights.

11 Like actions in trespass, riparian rights to water also display a remarkably formal rational character and it is notable that nothing equivalent to the locality rule in nuisance was developed by the courts in the nineteenth century in relation to water pollution

12 Although the tort of private nuisance has come to resemble the tort of negligence in certain respects (Gearty, 1989), the distinction between them in the meaning given to 'reasonableness' remains important, especially in the case of continuing nuisances (see Rogers, 1994 , p. 405).

13 This is not to suggest that 'private' rights should invariably prevail, only that the issue remains potentially open so that a case may be presented to a court.

14 Regulatory failure may take a variety of forms, including regulatory capture, overregulation to benefit larger firms by squeezing out competition and informational deficiencies. See generally Ogus, 1994.

15 If the affected householders in *Gillingham* had sued in private nuisance, then this issue could have been argued before the court . Why should the local authority bringing the action under local government legislation be in a different position? Arguably, the local council was acting not in its capacity as a planning authority but as a public health authority.

16 For example, in *Re Bowman* [1932] 2 KB 621, Swift J remarked (p. 633) that:

> [W]hen an owner of property against whom an order has been made under the Act comes into this court and complains that there has been some irregularity in the proceedings, and that he is not liable to have his property taken away, it is right, I think, that his case should be entertained sympathetically and that a statute under which he is being deprived of his rights to property should be construed strictly against the local authority and favourable towards the interests of the applicant, in as much as he for the benefit of the community is undoubtedly suffering a substantial loss, which in my view must not be inflicted upon him unless it is quite clear that Parliament has intended that it shall.

See generally Griffith, 1991, p. 114; Harlow and Rawlings, 1984, pp. 352ff.

17 Simon Ball has suggested that one approach to a permission is to make its effect on private rights subject to procedural safeguards: '[*I*]*f* there is compensation available, *if* there is a reasonable chance of a successful appeal and *if* there is less difficulty in bringing a judicial challenge, then the extinction of private rights by a public decision may be acceptable' (Ball, 1995, p. 296, original emphasis).

18 For a discussion (in a different context) of the way in which important cultural meanings can become attached to land, see John O'Neill, 'King Darius and the environmental economist', in this volume. We should also point out that we are not arguing that the status quo should always be preferred: our claim is that a space should be left to enable the (public interest) arguments for environmental conservation to be addressed through private legal action.

19 See, e.g., Mumma, 1992; the *Croyde Bay* case [1994] *Water Law* 183 The *Cambridge Water* case is an excellent example of private litigation's stimulating scrutiny of the effects of pollution, in that the case was responsible for important developments in the understanding of the movement of pollutants in aquifers (see *Allen v Gulf Oil Refining*, op. cit.).

20 An earlier version of this paper was given in the Kent Law School staff seminar series in March 1996. We are grateful to colleagues for their comments on that occasion and to Bill Howarth for reading and commenting upon a subsequent draft.

References

Ball, S. (1995), 'Nuisance and Planning Permission', *Journal of Environmental Law*, Vol. 7, No. 2, pp. 278–296.

Beckerman, W. (1990), *Pricing for Pollution*, 2nd edn, (Hobart Paper 66) IEA: London.

Brenner, J.F. (1974), 'Nuisance Law and the Industrial Revolution', *Journal of Legal Studies*, Vol. 3, pp. 403–433.

Brubaker, E. (1995), *Property Rights in the Defence of Nature*, Earthscan: London.

Coase, R. (1960) 'The Problem of Social Cost', *Journal of Law and Economics*, Vol. 3, pp. 1– 44.

Gearty, C. (1989), 'The Place of Private Nuisance in the Modern Law of Torts', *Cambridge Law Journal*, Vol. 48, Part 2, pp. 214–242.

Gray, J. (1994) 'Equitable Property', *Current Legal Problems*, Vol. 47, Part 2, pp. 157–214.

Greve, M.S. (1996), *The Demise of Environmentalism in American Law*, AEI Press: Washington, DC.

Griffith, J.A.G. (1991), *The Politics of the Judiciary*, 4th edn, Fontana: London.

Harlow, C. and Rawlings, R. (1984), *Law and Administration*, Weidenfeld and Nicolson: London.

Jacobs, M. (1991), *The Green Economy*, Pluto Press: London.

Karp, J.P. (1992–93), 'A Private Property Duty of Stewardship: Changing Our Land Ethic', *Environmental Law*, Vol. 23, pp. 735–762.

Maclaren, J.P.S. (1983), 'Nuisance Law and the Industrial Revolution – Some Lessons from Social History', *Oxford Journal of Legal Studies*, Vol. 3, pp. 155–221.

Mumma, A. (1992), 'Protection of the Water Environment Through Private Court Action', *Water Law*, Vol. 3, No. 2, pp. 51–54.

Nash, R. (1982), *Wilderness and the American Mind*, 3rd edn, Yale University Press: New Haven and London.

Ogus, A. (1994), *Regulation: Legal Form and Economic Theory*, Oxford University Press: Oxford.

Pearce, D., Markandya, A. and Barbier, E. (1989), *Blueprint for a Green Economy*, Earthscan: London.

Simpson, A.W.B. (1995), *Leading Cases in the Common Law*, Clarendon Press: Oxford.

Rogers, W.V.H. (1994), *Winfield and Jolowicz on Tort*, 14th edn, Sweet & Maxwell: London.

10 Unsustainable developments in lawmaking for environmental liability?

C.M.G. Himsworth

1 Introduction

If opportunities for the writing of history continue to be assured and the time does eventually come for an analysis of developments in environmental law in the closing years of the twentieth century, the record will probably reveal a series of turbulent interactions of events. Against what will appear to be a backdrop of accelerating environmental degradation, there is a mixed political and legal response. After some years in which governmental environmental activity has boomed – the United Nations Conference on Environment and Development at Rio in 1992, the expansion of EC environmental regulation in the Single European Act and the Maastricht Treaty, the UK Environmental Protection Act of 1990 and Environment Act 1995 and continuing developments in the United States all attest to this – we may now be seeing the beginning of a decline. Political change in the United States and in Europe may be producing a reversal or at least stagnation in environmental programmes.

Without the benefit of hindsight, it is difficult to characterise the state of the law in this transitional phase. Certainly, there is more of it than there once was. There are more practitioners engaged in environmental law, although probably, because of declining economic activity, not as many as might have been anticipated a few years ago. The UK statute book reflects the institutional initiatives which have brought the two new environmental agencies and also new or substantially revised regulatory schemes, including those designed to produce 'integrated pollution control'. In other respects, however, a more chaotic picture has developed – one which is characterised by the declaration of a boldly declared principle (or, to its detractors, a mere slogan) but then by the poor articulation of the principle in the legal regimes it supposedly informs. Chosen for examination in this paper is the 'polluter pays' principle and the ways in which it is reflected in the law of civil liability for environmental damage in the United Kingdom.[1] The principle that the polluter should pay for environmental damage is one represented at the highest level in EC law after its insertion into the Treaty by the Single European Act (Art. 130R (2))[2] and it has been claimed to be part of the UK Government's environmental policy and to have informed its legislative initiatives (see, e.g. *This Common Inheritance* Cm 1200, p. 13). It has been used, for instance, to justify the transfer to industrial polluters of the costs of regulatory institutions and procedures (including enforcement) (see Environmental Protection Act 1990, s.8 (Fees and charges for authorisations)). Centrally, however, it is supposed to inform the rules which impose a direct form of civil liability on all of those who have caused environmental damage and should pay compensation therefor.[3] Indeed, one might imagine that the principle would be represented at its strongest in this area. There is nothing quite so directly illustrative of the polluter's being made to pay as the imposition of an obligation to compensate in damages for harm caused. It will be suggested in this paper, however, that the ways in which a general liability of that sort are established in the law of the United Kingdom are at present very incoherent and poorly articulated. The rules and proposed rules with a little commentary are the subject-matter of section 2. In section 3, there is some discussion of the issues raised by this state of the law. In section 4 some conclusions are drawn. In particular, it will be argued that the pursuit of a single code of civil liability is misplaced and is to oversimplify issues which are too complex for such treatment. Running through the paper as a sub-theme is an argument that very little consideration indeed has so far been given to the incorporation into enforceable civil liability of ideas of environmental value and its quantification that have characterised the work of environmental economists and their critics. The question raised is whether environmental jurisprudence could or should accommodate not only the complexity of 'conventional' forms of damage and its quantification but also the expanded view of environment value required to include the valuation of a landscape destroyed, a species extinguished or intergenerational justice denied.

2   Current law: a brown-field site

The existing law of civil liability is a bit of a mess. It occupies a site which shows signs of earlier workings which have ceased to perform the function they once did and it is not clear whether they can adapt to modern conditions. There are also sporadic instances of more recent development, but their application is limited and it is not clear whether their technology is of good quality. There are proposals for the overall regeneration of the area by new laws of general application, but their viability too is in question.

In the UK, legal systems one would expect regulation in this area to be the product either of statutory intervention or of rules of the common law, or a mixture of the two. Increasingly, one has also come to expect that any changes in the law would be EC-driven. In this account, there is first a brief summary of relevant statutory provisions. Then – because if they were to be brought to implementation in the United Kingdom it would be by statutory means – the current European developments and proposals are considered. The EC Commission's Green Paper may yet turn out to be the more significant, but here it is preceded by mention of the Council of Europe Convention. Finally, common law developments are considered.

As far as the current statute book is concerned, there is, perhaps surprisingly, only sporadic direct provision for civil liability for environmental damage.[4] There is, for instance, important provision for liability arising out of nuclear installations (Nuclear Installations Act 1965 s.12) and radioactive substances (radioactive Substances Act 1993 s.30). Another specific provision is s.73(6)[5] of the Environmental Protection Act 1990 which appears in the part of the Act dealing with the disposal of waste. It refers to other provisions within the Act which impose criminal liability for the unlawful disposal of waste on land and then attaches civil liability in the same circumstances. Where any damage[6] is caused by waste which has been deposited in or on land, any person who deposited it, or knowingly caused or knowingly permitted it to be deposited, in either case so as to commit an offence (under s.33(1) or s.63(2) of the Act), is liable for the damage.[7] Exceptions to this general rule are where either the damage was due wholly to the fault[8] of the person who suffered it or the damage was suffered by a person who voluntarily accepted the risk of damage. The liability created by virtue of s.73(6) is stated to be without prejudice to any liability otherwise arising.

Although this section is more broadly cast than its predecessor provision (Control of Pollution Act 1974 s.88), it is far from clear what it can be expected to achieve. It is technically difficult to construe because of its interconnection with the separate provisions establishing criminal liability (e.g., the defences available in a criminal prosecution are stated also to be available in the civil proceedings (s.73(7)) but it also leaves much unsaid. A person is declared to be 'liable for the damage' but it is nowhere made clear to whom he or she is liable. Section 33(1), which creates one of the criminal offences from which civil liability derives, forbids anyone from inter alia disposing of waste in a manner likely to cause 'pollution of the environment or harm to human health'. That may be a formula sufficient to create and define a criminal offence but it is not one which identifies to whom the person who deposits waste may be civilly liable. If the 'environment'[9] is polluted[10] it is far from clear who is permitted to sue the alleged polluter. Nor is it at all clear how the liability should be quantified. Damage is defined to include death of or injury to persons and may be assumed to extend to property damage but is not confined to those categories (Ball and Bell, 1995) and there seems to be no reasonable way, in reliance upon the text of the Act alone, to guess what parameters should be placed around the damages which may be claimed. There is no indication of whether losses described in the context of tort law as 'purely economic' – as opposed to losses associated also with injury to person or property – would be compensatable or not. Presumably they would be. And, as one moves (as is presumably permitted under the section) from damage to the 'owned' to damage to the 'unowned' environment, what value is to be attributed to losses there? How far must assessment of loss be confined to damage to which a commercial value can be given? Can the section be used to compensate losses of non-commercial 'users' of the environment, i.e., those who do not depend on the environment for their earnings but who do, for instance, use the environment for amenity purposes. Sports fishermen, scuba divers and bird watchers have been suggested as examples in this category.[11] Thereafter there are what have been called 'non-use' values: 'option value', representing a general desire to preserve the option to use a natural resource in the future; 'existence value', which reflects simply the knowledge that the resource exists; and the 'bequest value' that derives from the knowledge that it will be available to future generations. The mere listing of these categories of 'environmental value' is sufficient to expose the complexity of the quantification of liability in this area. It is misleading, however, if it gives the impression that simple choices may then be made about the categories of loss to be incorporated into a rule of liability. Above all, that would beg the question of the reducibility of the wider categories of environmental value by some process of cost benefit analysis to values quantifiable in money terms – a process which would presumably be an essential prerequisite of the calculation of an award of damages following legal action but a process whose validity is substantially denied both by the critics of any use for this purpose of cost benefit analysis (see, inter alia, Sagoff, 1988 and Hayward, 1994) and even by the proponents of cost benefit analysis themselves.

It may be responded that to elaborate these difficulties of identification and quantification of damage which may attract liability is to overstate the problems which are actually likely to arise in the interpretation of a statutory provision of apparently narrower compass. Perhaps this is so, but it does enable us to appreciate the range of considerations which might need to be taken into account in the setting up of a more general statutory scheme of liability.

Another series of statutory provisions, also confined to a fairly specific sector of damaging activity, is that recently inserted as Part IIA of the 1990 Act by the Environment Act 1995 and which deals with the 'remediation' of contaminated land. The complex new code will require local authorities to identify contaminated land and then (in some cases involving one of the Agencies instead as the enforcing authority) to serve on one or more 'appropriate persons' a remediation notice specifying works to be undertaken. An appropriate person is, in the first instance, a person 'who caused or knowingly permitted the substances, or any of the substances, by reason of which the contaminated land in question is such land to be in, on or under that land'(1990 Act s.78F (2)) but 'appropriateness' may also extend to the owner or occupier for the time being of the contaminated land (s.78F (4)–(5)).[12] There is provision for appeal against a remediation notice (s.78L) but then noncompliance without reasonable excuse is a criminal offence (s.78M) and, most importantly to this account, the enforcing authority has powers to carry out remedial works in situations including those where the appropriate person has failed to comply with a notice and then to recover the cost of that remediation (ss.78N–78P). This model of imposition of liability for damage caused or threatened by means of recovery of cleanup costs in default of cleanup by the polluter is one to whose advantages and disadvantages we should return.[13]

A second legal approach to civil liability comes in provisions potentially of application across Europe. It throws together the distinct styles and status of (a) a Council of Europe Convention (the Lugano Convention of 1993) and (b) the proposals under discussion within the European Community for general rules on civil liability for environmental damage. The Lugano Convention on Civil Liability for Damage Resulting from Activities Dangerous to the Environment, to which the United Kingdom has not acceded and which, even if it were to accede, would have only the status of a treaty binding on the state and would not carry with it the requirement of direct application by the courts familiar in the case of EC law, 'aims at ensuring adequate compensation for damage resulting from activities dangerous to the environment and also provides for means of prevention and reinstatement' (Art. 1). It does so, having regard, inter alia, to 'the desirability of providing for strict liability in this field taking into account the "Polluter Pays Principle"'. Despite its lack of application in the United Kingdom,[14] the main interest provided by the Convention probably lies in its ambition not only to apply to many different state parties but then also to a very broad range of circumstances within those states instead of restricting its application specifically to particular, identified situations. This, nevertheless, forces the need to define the areas to which it does apply, which is done by means of definitions of e.g., 'dangerous activity', 'damage' and especially 'environment', which is defined very broadly to include:

– natural resources both abiotic and biotic, such as air, water, soil, fauna and flora and the interaction between the same factors;
– property which forms part of the cultural heritage; and
– the characteristic aspects of the landscape.

Liability falls on an 'operator' – 'the person who exercises the control of a dangerous activity' and, subject to certain exemptions, liability is 'strict'. However, although there is some reference in the Convention to a person who has suffered damage and there is an implication that such a person is the one who might raise an action under the Convention's terms, there is no formal identification of whom may be regarded as having suffered damage – who, for instance, suffers where damage is done to the 'characteristic aspects of the landscape'? – nor of how liability is to be measured.

A feature of the EC Commission's Green Paper on 'Remedying Environmental Damage' (COM(93) 47)[15] and the responses to it has been their attempts to grapple with the difficulties (both conceptual and political) of putting some definitional limits on the idea of a generalised civil liability for damage: liability for what kinds of damage to what defined aspects of the 'environment' and with fault-based or strict liability; the problem of damage arising from past pollution; the sharing (and channelling) of liability between those held liable; who may sue and with what remedies in view; and the impact of insurance on systems of liability. Another idea floated by the Green Paper was recourse to joint compensation schemes with payments made to those suffering damage from funds contributed from relevant economic sectors. Progress towards the preparation of firmer proposals on liability as a precursor to implementation by directive has not been untroubled. Indications at the time of writing are were that, despite some signs of a wish by the environmental commissioner to see progress, the adoption of proposals into the Commission's legislative programme is not imminent ((1995) ENDS Report 251, p. 35).

There has, in the meantime, been a flurry of interest in the capacity of the common law, unaided by direct statutory input, to provide remedies for environmental damage. The tort of nuisance has historically (in the days preceding adoption of the terminology of 'environmental' law and liability) been viewed as the most relevant (see, e.g., McLaren, 193) and, with the addition of the strict liability regime contributed by the rule in *Rylands v Fletcher* ((1866) LR 1 Ex 265; affirmed (1868) LR 3 HL 330),[16] most promising today as a source of general principles of liability. Recent attention has focused most strongly on the case of *Cambridge Water Co Ltd v Eastern Counties Leather plc* ([1994] 2 AC 264) in which the House of Lords, overturning the Court of Appeal, held the defendants not liable under the *Rylands v Fletcher* principle for damage caused to a source of water supply as a result of seepage of a chemical through the aquifer over many years eventually producing contamination to a level forbidden by an EC drinking water directive. The offending tannery's claim to be protected from liability by being a 'natural user' of the land was rejected but there would be no liability in circumstances where, if an escape were (however improbably) to take place, the damage such an escape caused was at the time of the escape unforeseeable.[17] The case has been viewed as a setback for environmentalists, as commercial good sense by industrialists and as providing a significant contribution to the development of tort law by commentators in that field. We shall return to the question of how far the case has provided further insights into the use which may be made in future of common law tort rules in the area of environmental liability, whether or not in conjunction with rules from other sources.

3   Contaminated law?

A leading House of Lords decision which absolves from civil liability those who have polluted drinking water, an unratified international treaty, thwarted attempts to make progress towards an EC directive and sporadic, little-used and perhaps unintelligible provisions on the domestic statute book may not be the cause of much surprise. To many observers this would be the entirely predictable result of any attempt to impose by legal regulation any sort of restrictions on commercial profits which liability for environmental damage could entail. Without, however, making overambitious and unjustifiable assumptions about the capacity of legal regulation to induce change in the commercial world, it seems not unreasonable to suppose that some legal strategies will make more of an impact than some other legal strategies and it is valuable to think carefully about the options available. Meanwhile, different judgements will be made about what should be the goals sought by environmental policy and law, informed in some cases by quite different assumptions about the ordering of national and international economies and societies and different judgements will be made about the more technical issues concerning how they may be attained. These are arguments which rage fiercely across the United States, where governmental environmental intervention especially by means of environmental law has, in the modern era, been longer established and its impact, if not better understood, more debated.[18]

The question about the use to be made of environmental law divides those whose preference is for strong governmental intervention from those whose instinct or political commitment or experience is against it. It is plain that those who proclaim a sympathy for broad environmental objectives but who dislike the prospect or present reality of Madison's nightmare[19] of an over-powerful central government will have little truck with the 'command and control' forms of governmental intervention by strong and, some would say, oppressive agencies of the state. Despite the certain worsening of environmental conditions nationally within the United States and globally and the apparent strengthening of the case for tighter governmental regulation, the experience of the Superfund,[20] in particular, has done more than merely

167

confirm the suspicions of those already sceptical of big government but has gone as far as to create a consensus across a much wider range of opinion that the machinery for imposing cleanup liability was signally unsuccessful. Instead of 'command and control', government sceptics offer measures more sympathetic to the operation of the market and more likely to be favourably regarded by industry. Tax or grant incentives are the archetypal instance. As applied more specifically to issues of environmental liability, this schema readily divides those who would use a strong role for governmental institutions' imposing cleanup costs on miscreant industrialists from those would prefer a softer, more market-orientated approach which, in this instance, probably presents as a preference for the operation of the common law and the resolution of disputes by civil proceedings in the courts.

There is another dividing line which runs through the environmental policy debate at almost the same place, i.e., between the use of development of the common law and the promulgation of new statute law, but supported by a rather different rationale. There are, it is said, many difficulties in the way of 'using' the common law to achieve particular social ends. It is readily conceded that courts do, through their decisions, make new law and, though the means of doing so may be rather different from that of a legislature, the impact of deciding a case one way or the other may be just as great. There is no doubting the impact, for instance, on the parties' pockets in *Cambridge Water* when the House of Lords took a different view of the law from the Court of Appeal – the Court of Appeal had awarded over £1m in damages. But this itself illustrates what many take to be the lottery of the common law. Judges do probably seek to base their decisions on principles and probably it is a good thing that they do[21] but their principles fluctuate and there are doubts about the capacity of judges to integrate the higher arts of, e.g., cost-benefit analysis into their formulation of legal principle. In the development of principles in the field of environmental liability, it has been recognised that, in contrast with other areas, there are particular problems. They have been characterised as 'problems of identification' (of, e.g., a pollutant in the atmosphere), 'problems of source' (where it came from and how to prove it), 'problems of boundaries' (especially the extent of the population or environment affected) and 'problems of common interests' (the problems already noted of how to take account of heritage and global issues).[22] Judges depend furthermore on the chance events of cases being brought before them. The terms in which they must frame their conceptual distinctions are more confined and their vocabulary narrower, than is the case with legislatures. As Lord Goff said in *Cambridge Water*, 'statute can where appropriate lay down precise criteria establishing the incidence and scope of such liability' ([1994] 2 AC 264, p. 305) – although it has to be admitted that the opportunity to lay down precise criteria is not always taken, as arguably was the case with s.73 (6) of the Environmental Protection Act 1990 discussed above. It is also recognised that it is the task of legislatures on the whole, to make general laws proactively rather than, as is the case with courts, to make quite specific decisions retroactively. Above all, it is said, however much courts may be necessary to the process of dispute resolution, they lack the democratic mandate of a legislature to make new law.[23] The wish of judges to defer to the proper role of parliament to develop the law strategically in ways which it sees fit is commonly expressed in the law reports and did, for instance, make an appearance in *Cambridge Water*, where Lord Goff: said that 'it appears that, under the current philosophy, it is not envisaged that statutory liability should be imposed for historic pollution'; cited the draft Council of Europe convention in support; and concluded that, if this is so, 'it would be strange if liability for such pollution were to arise under a principle of common law' ([1994] 2 AC).[24] These differences between common law development and the making of statute law should not be overstated (just as reliance should not be unwisely placed, for that matter, on the divide which is sometimes recognised in this area between 'private' and 'public' law) but they do have some relevance to the debate, if only as a reminder that in most fields the total picture of the law consists of a mix of the two and that it is foolish to consider a policy area as a whole without (eventually) taking account of both.

It will, however, be one of the conclusions to be drawn in this paper that common law liability for environmental damage will not play a large part in the future development of patterns of liability as a whole. On the other hand, it would be foolish to ignore vibrant debates within tort law at large in so far as they may have a bearing either upon environmental common law itself or the ways in which common law doctrine may be adopted into statutory schemes.

Controversy around the purpose and function of tort law (and especially its most important characteristic as a system of law designed to offer plaintiffs the prospect of full compensation, if successful) is not new, although increasing attention has probably been given to it in recent years, partly because of an increase in numbers of (arguably) unprincipled decisions by courts producing indefensible distinctions between those individuals who recover damages in tort and those who do not,[25] but partly reflecting new concerns about what should be the relationship between the burdens of compensation imposed by tort law on individual defendants and, on the other hand, the burdens which are 'collectivised' by being spread across a wider population either by schemes of insurance or by schemes of state provision for damage and loss incurred.[26] There can be no doubt that some of this more general debate has a close bearing on nuisance law and its relevance to environmental issues. Plainly, with so much at stake, there is a concern, all other developments in the law apart, about the capacity of the law of nuisance to provide a principled basis on which findings of liability can be grounded and a degree of certainty and predictability achieved. The array of opinion at the different levels of *Cambridge Water* (quite apart from the range of opinion in academic commentary) is not promising. There is no doubt too that environmental liability based on tort must be considered against the rival possibilities of the spreading of risk by means of insurance or state-imposed systems of compensation. The EC Green Paper reflects these considerations.

There are, on the other hand, good reasons for not pressing too far the parallels between arguments relevant to the resolution of wider problems in the law of tort and arguments of more specific relevance to environmental liability. Thus it is probably true to say that in the law of tort at large, dominated as it is by principles in the law of negligence, the primary concern is with the provision of full (or, at least, adequate) compensation to individual plaintiffs who have been injured or become ill as a result of an 'accident' whether on the roads or in the workplace or in countless other situations. Of course, it will also be argued, with greater or lesser force according to the particular situation, that the law can or should be able to perform the more symbolic role of deterrence of behaviour thought socially undesirable. It would be a good idea if the law not only compensated victims but also punished and deterred careless employers and drivers and this becomes an obviously relevant consideration at the point where alternative means of compensating victims from sources other than the pockets of individual wrongdoers are considered and especially where these are perceived to fail, through the dilution of risk and liability, to provide the same measure of deterrence. Also relevant are the problematic issues of the impact of socialisation of risk upon the 'autonomy' of individual victims and the creation of a culture of 'dependency', as is warned by some pro-tort, anti-collectivist commentators.[27]

Two comments may be offered on this. The first is that the very existence of this debate is a useful reminder that these questions are political questions (and sometimes more political than they seem at first to be); they involve judgements to be made about the spreading of risk and about the role of the state and the law in the spreading of risk which go far beyond issues of a technical legal nature and which almost certainly should be assigned to legislatures, rather than to courts, for resolution; and without doubt, these judgements are as relevant to environmental liability as they are to other forms of accident liability. On the other hand, it also seems clear that some of the preoccupations of mainstream tort lawyers are not readily applicable to the interaction of nuisance law and environmental protection. This is because the concern of environmental lawyers is with the environment. Theirs is not an ancillary, symbolic concern to protect the environment by deterring damage which is ranked in second place after the primary concern of compensating individual victims. Protecting the environment is the primary aim. This is not to say that there are not victims of environmental damage but, especially if the focus is on nuisance, the environmental concern is not primarily to ensure full compensation for their loss. It may seem a harsh thing to say about the Cambridge Water Company but the environmental point was not to reward them. Rather it was to produce a decision which would 'make the polluter pay' and act as a deterrent to polluters at large. In so far as the question of compensation arises, the aim is somehow to ensure that it is the 'environment' that is compensated. Little importance is attached to the reimbursement of the Cambridge Water Company, to the protection or not of its 'autonomy'; or to preventing it from falling into a condition of dependency! If the law of nuisance can be made to serve the broader environmental purpose, well and good. If, on the other hand, despite its apparently long-established environmental credentials, it turns out to offer no more than a system which is an adjunct to the general law of property and which merely serves as a protection for the rights of property owners inter se – a process easily producing litigation which, whilst of potential private benefit, runs directly against the wider environmental interest – then it must be set on one side. To conclude otherwise would be to attach oneself to a deeply conservative, heavily property-based and widely rejected theory of environmental law. On any more enlightened view, the law of nuisance invoked necessarily by individuals will be welcome only if those individuals are capable of acting as efficient surrogates of a wider interest. Happily this caution about making the common law work too hard for the environment is reciprocated in the views of those whose principal task it is to develop the common law. Lord Goff's concern in *Cambridge Water* was not simply to defer to the greater efficiency and legitimacy of legislation in defining new rules of environmental liability, but was also to protect general common law principles, for which he was seeking to establish a coherent basis, from the distortion which might result from trying to serve the single issue of environmental pollution ([1994] 2 AC 264, p. 305).

4   Conclusions: law fit for the purpose

These thoughts about the limitations of the law of nuisance in any environmental strategy – limitations which, whilst not prompting the necessity to intervene by statute to the extent of abolishing existing rights and remedies (let those who are able to found neighbourly disputes on nuisance continue to do so), do appear to leave tort law as a very low-ranking priority on the environmental lawyer's agenda – serve to prompt concluding reflections on how best to ensure that rules of civil liability give some renewed recognition to the principle that the polluter must pay.

1   What certainly emerges is an understanding that the notion of a general 'civil liability for environmental damage' is altogether too wide for

sensible discussion. Without further refinement, it fails to identify what is meant by the 'environment' and 'damage'; who may be held liable for the damage; how, and in the light of what criteria, damage may be quantified; and, perhaps most important of all, who may be entitled to sue in respect of the damage and who should be the recipient of any compensation awarded. Not enough has been said in this paper about these last points but they are, as already noted, insufficiently addressed in most existing rules on liability. If the 'environment' is to be broadly defined, as, for example, in the Lugano Convention, it should follow that the rules identifying potential plaintiffs and recipients of compensation will reflect that breadth. The measure of harm has to be capable of taking account of not only of damage done to the person and property of individual human beings, but also perhaps of damage to future generations of human beings and other flora, fauna (including their diversity) and aspects of cultural heritage and aesthetics of the natural environment which are, in the ordinary way, only with difficulty accorded standing or designated as the beneficiaries of compensatory awards. What this tends to point to is the starting assumption that any award for environmental (as opposed to merely private) injury should be channelled in the direction of some public fund – the same fund as that into which criminal fines should be paid and the same funding source that may be available to compensate for development restrained in the name of conserving the environment. Typically and inevitably, of course, this will mean the state and it is a matter of interest that some statutory schemes of environmental liability do give direct recognition to this. In the relevant Italian legislation, for instance, this seems to be the case (see Bianchi, 1994; Biamonti, 1994). However, this does not necessarily imply that, though the state be the only (or main) beneficiary, the state should be the only plaintiff and it is, in this respect, instructive to take account of the opportunity given by the Lugano Convention to 'appropriate organisations' to initiate litigation. Another illustration is where, under the US Superfund legislation, 'natural resource damage as (for injury to the unowned environment) may be recovered by government trustees of the resources concerned (see Stewart, 1995).

2   There is an inevitability here that it should be the state by legislation which establishes a scheme or schemes rather than judges by development of the common law – bearing in mind, equally inevitably, that this may yet have to be done in accordance with instructions given by Directive from Brussels. The long delays in the process have been noted and have, for some, rekindled a preference for resort to common law solutions (see Ogus, 1994) but it would be impossible for these to provide the radical changes necessary.

3   It has also become plain that any strategy embarked on in relation to 'civil liability' should be undertaken, and preferably in a coordinated way, in full recognition of the possibilities available on other fronts.[28] Because of their institutional integration into the powers available to their Environmental Protection Agency, the parallels between taking civil action, undertaking a criminal prosecution or securing the carrying out of environmental good works in lieu are better recognised in the United States than in the United Kingdom. Similarly, some European countries (e.g., the Netherlands) operate systems of liability which include the 'administrative fine'.[29] The necessary links to be made – between remedial measures, in the form of civil liability and damages and, on the other hand, the serious adoption of incentive-based measures as well as continued reliance on 'command and control' measures for prevention of damage in the first place – are already more evident. And an integral part of any reforming programme is ensuring that all relevant parties have sufficient access to the information necessary to support or repel a legal claim (especially that gathered under statutory authority by public bodies).

4   Perhaps the question most poignantly raised is whether – in the light of the problems involved in defining civil liability and the need that, in whatever version or versions it is adopted, a concept of liability should be carefully integrated into surrounding patterns of legal regulation – the idea of a single all-embracing doctrine of liability is at all tenable. It has to be the provisional conclusion of this paper that it is not. This is not the same as the argument that, whether it be attempted by Council of Europe Convention or by EC Commission Directive, harmonisation of the law across European states is not desirable or achievable, although such a scepticism has indeed informed the approach of the British government so far. Nor, one hopes, is this an argument which would in all circumstances reject what may be categorised as the civilian preference for the articulation of the law in broad principles in favour of some sort of common law pragmatism. Rather it is to argue that, in this instance, a project which seeks to embrace a vast range of possible instances within a single doctrine of environmental liability is quite meaningless. The fact that there *is* such a wide range of types of damage which may be taken under review and then that there is a wide range of suggestions of legal strategies and eventually of the more specific rules required is fully apparent in the evidence submitted to the enquiry conducted by a subcommittee of the House of Lords European Communities Committee and which reported in 1993 in *Remedying Environment Damage* ((1993–94) HL Paper 10). The focus of the enquiry was the EC's Green Paper and both the evidence gathered and the Committee's own conclusions

contain much of interest. The evidence reveals (as do, in some measure, the Committee's recommendations) two main problems. The first is that of definition already noted in this paper. There is simply no consensus on what it means to impose civil liability for damage. In particular, some see it as a mechanism substantially unrelated to wider strategies of remediation achieved, for instance, by compulsory cleanup and recovery of expenses, whereas others insist on seeing a much closer relationship between the two. The other main phenomenon is that of the piecemeal approach to reform. Towards the environmental campaigning end of the spectrum, it is, for instance, not surprising to find strong demands for strict rather than fault-based liability and for generous rules on access to the courts for litigants (replicating, in this case, the claims for a broad interpretation of *locus standi* in judicial review where the issues may be quite different) but without recognising the obligation to integrate issues of strictness of liability or of access into a comprehensive scheme of liability and compensation. It is when all the threads are drawn together, as was attempted by the House of Lords Committee, and the discipline is imposed of weighing all issues in the balance *all* the time that the enormity of the task and the absence of a sufficiently focused organising ideal is revealed. Certainly the radical vision of a world in which polluters, however they are defined, do indeed pay can be retained. But the route to the attainment of that ideal is best seen in terms of the modest incrementalism of devising specific rules for specific problems in a coordinated way. In an EC context in which the specific rules, their coordination and enforcement would be the responsibility of member states subject to guidance at a Community level, this presents a special challenge.

Notes

1 Another such principle which might be considered in a similar way is the 'precautionary principle'. See O'Riordan and Cameron, 1994.
2 'Community policy on the environment shall aim at a high level of protection taking into account the diversity of situations in the various regions of the community. It shall be based on the precautionary principle and on the principles that preventative action should be taken, that environmental damage should as a priority be rectified at source and that the polluter should pay'.
3 On the related issue of compensation for state restrictions imposed on environmental grounds, see Rowan-Robinson and Ross, 1993.
4 For a useful summary of the provisions applicable in England and Wales, see Burnett-Hall, 1995, pp. 943–6.

5   In force since 1 May 1994.

6   Defined to include the death of, or injury to, any person (including any disease and any impairment of physical or mental condition) (s.78(8)).

7   For the purposes of the Fatal Accidents Act 1976, the Law Reform (Contributory Negligence) Act 1945 and the Limitation Act 1980 and also for the purposes of any action of damages in Scotland arising out of the death of, or personal injury to, any person, any damage for which a person is liable under s.73(6) is to be treated as due to his fault (s.73(9)).

8   Defined as having the same meaning as in the Law Reform (Contributory Negligence) Act 1945.

9   Defined in Part II of the Act as consisting of 'all, or any, of the following media, namely land, water and the air' (s.29(1)).

10   'Pollution of the environment' is defined to include the escape from land of waste which is capable of causing 'harm' to man or any other living organisms supported by the environment (s.29(2)). 'Harm' is defined to include, in the case of man 'offence to any of his senses or harm to his property' (s.29(5)).

11   For the general typology of ch. 3 environmental value relied on here see Pearce et al, 1989 and subsequent works. See also para. 18 of the Memorandum of the International Tanker Owners Pollution Federation Ltd submitted to the House of Lords Select Committee on the European Communities and published with the Committee's report *Remedying Environment Damage* (1993–94) HL Paper 10, p. 145.

12   Further rules enable the Secretary of State to determine, by guidance issued under the section, which of several persons should be treated as appropriate (ss.78F (6)–(8) and 78YA).

13   Other provisions on this model are the Water Resources Act 1991 s.161 (E and W) and (for Scotland) the Control of Pollution Act 1974 s.46 – as inserted by the Water Act 1989.

14   Or indeed, anywhere else so far.

15   The proposals in this Green Paper are seen as overtaking the Commission's earlier proposal for a Directive on Civil Liability for Damage Caused by Waste [1991] OJ C192. That draft Directive was more limited in that it was confined to waste, but it was a document which necessarily addressed the definitional problems concerning, e.g., 'damage' and 'impairment of the environment' and the issue of who may be a plaintiff in proceedings under the Directive.

16   In Scotland, there are strong parallels recognised between delictual liability in nuisance and the English law but it has been said, most forcefully in *RHM Bakeries (Scotland) Ltd v Strathclyde Regional Council* 1985 SLT 214, that the rule in *Rylands v Fletcher* does not apply.

17   For comment on the case, see, inter alia, Weir, 1994a and Ogus, 1994.

18 A review of developments discussed at a Symposium on 'The Environment and the Law' was published at (1994) 21 *Ecology Law Quarterly* 329.
19 The phrase used by Stewart in his 'Antidotes for the "American Disease"', 1993.
20 Established under the Comprehensive Environmental Response, Compensation, and Liability Act 1980 (CERCLA).
21 Though on this see Weir, 1994b.
22 See Brüggemeier, 1994, in reliance upon Rabin, 1987. Brüggemeier's paper provides an excellent account of some of the issues we have sought to address in this paper.
23 For a useful brief summary of the arguments, see Sunstein, 1994.
24 For a criticism of this approach, see Robinson, 1995.
25 For a discussion of this and the suggestion of a strategy for courts to 'pocket' tort issues in a more principled manner see Stapleton, 1994 and, for a rebuff, Weir, 1994b.
26 See e.g. the works of Jane Stapleton, most recently her 'Tort, Insurance and Ideology', 1995.
27 For discussion of this, see Stapleton, 1995.
28 For an excellent review of future possibilities especially as they relate to 'mass torts' and the integration of liability and regulatory standards, see Schuck, 1986.
29 For a full review, see the McKenna Report for the EC Commission, 1996.

References

Ball, S. and Bell, S. (1995), *Environmental Law*, 3rd edn, Blackstone: London.
Biamonti, L. (1994), 'Italy' in D. Campbell (ed.), *Environmental Hazards and Duties of Disclosure*, Graham and Trotman: London.
Bianchi, A. (1994), 'The Harmonization of Laws on Liability for Environmental Damage in Europe: an Italian Perspective', *Journal of Environmental Law*, Vol. 6, No. 21.
Brüggemeier, G. (1994), 'Enterprise Liability for Environmental Damage: German and European Law' in G Teubner et al. (eds), *Environmental Law and Ecological Responsibility*, Wiley: Chichester.
Burnett-Hall, R. (1995), *Environmental Law*, Sweet and Maxwell: London.
Hayward, T. (1994), *Ecological Thought*, Polity: Cambridge.
McKenna Report for the EC Commission (1996), *Study of Civil Liability Systems for Remedying Environmental Damage*.
McLaren, J.P.S. (1993), 'Nuisance Law and the Industrial Revolution', *Oxford Journal of Legal Studies*, Vol. 3, No. 155.

Ogus, A. (1994), *Journal of Environmental Law*, Vol. 6, No. 151.

O'Riordan, T. and Cameron, J. (eds) (1994), *Interpreting the Precautionary Principle*, Earthscan: London.

Pearce, D. et al. (1989), *Blueprint for a Green Economy*, Earthscan: London.

Rabin, R.L. (1987), 'Environmental Liability and the Tort System', *Houston Law Review*, Vol. 24, No. 27.

Robinson, D. (1995), 'Public Interest Environmental Law – Commentary and Analysis' in D. Robinson and J. Dunkley (eds), *Public Interest Perspectives in Environmental Law*, Wiley Chancery: Chichester.

Rowan-Robinson, J. and Ross, A. (1993), 'Compensation for Environmental Protection in Britain: a Legislative Lottery', *Journal of Environmental Law*, Vol. 5, No. 245.

Sagoff, M. (1988), *The Economy of the Earth*, Cambridge University Press: Cambridge.

Schuck, P.H. (1986), *Agent Orange on Trial*, Harvard University Press: Cambridge, Mass.

Stapleton, J. (1994), 'In Restraint of Tort' in P.B.H. Birks (ed.), *The Frontiers of Liability*, Vol. 2, Oxford University Press: Oxford.

Stewart, R.B. (1993), 'Antidotes for the "American Disease"', *Ecology Law Quarterly*, Vol. 21, No. 85.

Stewart, R.B. (1995), 'Liability for National Resource: Beyond Tort' in R.L. Revesz and R.B. Stewart (eds), *Analyzing Superfund: Economic, Science and Law*, Resources for the Future: Washington DC.

Sunstein, C.R. (1994), 'Public versus Private Environmental Regulation', *Ecology Law Quarterly*, Vol. 21, No. 455

Weir, T. (1994a), '*Rylands v Fletcher* Reconsidered', *Cambridge Law Journal*, 216.

Weir, T. (1994b), 'Errare humanum est' in P.B.H. Birks (ed.), op. cit.

11 Cultural communities and intellectual property rights in plant genetic resources

Anthony Stenson and Tim Gray

1   Introduction

This paper examines and rejects the view that cultural communities are morally entitled to intellectual property rights (IPRs) and their associated knowledge in plant genetic resources (PGRs). Firstly, then, we should define IPRs and the concept of moral entitlement. We take intellectual property rights to refer to any legal recognition that a creation or resource is the intellectual property of a specified individual or group. It is important to note that 'IPRs' is not a synonym for 'farmers' rights' (Crucible Group, 1994, pp. 34ff.; Brush, 1996, p. 139). The conception of farmers' rights adopted by the Food and Agriculture Organisation would oblige countries with plant breeding industries to pay money into a central fund, which would then be used to set up in situ biodiversity conservation programmes (FAO, 1989, p. 13, n. 1). Farmers' rights, then, as Brush points out (1993, p. 662), recognise the ongoing contributions of cultural communities' farmers *without* directly attributing ownership to, or commercialising, their resources and knowledge.

A moral entitlement theory of property is one that sees property as a primary moral right that is not justified by reference to consequences. Entitlement theories attribute rights in specific things to specific individuals or groups on the basis of a special relationship between object and claimant. Utilitarian theories that justify property by arguing that it promotes society-wide efficiency are not, therefore, moral entitlement theories; neither are those that argue that property is necessary to ensure peoples' ability to express themselves. The labour theory of property, which claims that ownership in something arises as a result of working on that thing, is an entitlement theory: to have laboured on something is the special relationship that gives rise to the moral entitlement. Consequences are irrelevant.

Certain kinds of IPRs have in fact been granted to cultural communities in PGRs. Recently, for example, the Kani tribe of the Indian state of Kerala was granted IPRs in the active ingredient of a plant-derived drug named *jeevani*, which is said to combat stress. The Kani received a $25 000 'know-how' fee and will also gain a share of the two per cent royalty on any future sales of a medicine developed from the plant (Jayaraman, 1996). These IPRs do not fall into any of the standard Western categories of IPRs – patents, copyrights, trademarks and trade secrets – but we would define them as IPRs because they recognise a specified community's ownership rights in a specified substance.

Arrangements like this have backing in the international arena. The annexes to the Food and Agriculture Organisation's 'Undertaking' on PGRs, agreed in 1989 and 1991, which affirm the principle of 'farmers' rights', formed the basis of certain clauses in the 1992 Convention on Biological Diversity. Article 8(j) of the Convention extends the principle to indigenous communities (UNEP, 1992). It has been argued that this clause can act as the basis in international law for granting to cultural communities full IPRs in PGRs (Da Carta E Silva, 1995, p. 546). The Kani agreement suggests that this might be happening.

There are many consequentialist arguments in favour of granting IPRs to communities. They might be necessary to conserve biodiversity (Margulies, 1993), for example, or to protect traditional ways of life (Greaves, 1995, p. 203). However, this paper is not concerned with these consequential questions. It is intended as a rebuttal of only one argument: the idea that cultural communities hold moral entitlements to IPRs in PGRs. It is not an attack on all possible formulations of 'farmers' rights', nor on the idea that there is something unjust about present arrangements. Neither is it a defence of Western transnational corporations and their governments. It is not even intended to refute altogether the idea that IPRs should be granted to communities over PGRs; just that an entitlement theory cannot justify this practice.

Some commentators argue that the treatment of 'raw' germ plasm as an unowned resource, and therefore freely available,[1] by the plant breeders of the 'North', is robbery and that the subsequent protection of 'elite' varieties of plants by these breeders is rank hypocrisy, adding insult to injury:

> ... in manipulating life-forms you do not start from nothing, but from other life forms *which belong to others* ... Third World countries [are] the original owners of the germplasm ... wild material is owned by sovereign states and by local people (Shiva, 1991, pp. 2745–6, italics added).

These writers have argued that if First World innovators are entitled as of right to patents on their inventions, then so are Third World communities and countries entitled as of right to some form of IPRs in their landraces,[2] wild genetic resources and botanical knowledge.

This position takes for granted an entitlement theory of intellectual property. We argue, however, that from the point of view of such a theory, there is no hypocrisy involved in asserting IPRs for special genetic stocks, yet denying them for original germ plasm; indeed, an entitlement theory cannot ground IPRs for communities in any sort of genetic resources. If Third World campaigners wish to argue for IPRs for cultural communities, they should abandon the idea that these communities are entitled *by right* to intellectual property over genetic resources and instead concentrate on consequentialist arguments.

We examine two types of entitlement claim made by advocates of IPRs in PGRs for cultural and local communities: the first is based on a labour theory of ownership; the second, on the idea that the proximity of a people to a genetic resource creates an IPR in it. We reject both theories. In the third section, we discuss, and reject, the idea that communities hold intellectual property entitlements to their folk knowledge.

2   The community labour theory

On the labour theory of intellectual property (IP) ownership, it could be argued that agricultural communities own their landraces by virtue of the fact that they have created them through centuries of seed selection and innovation. In claiming landraces as common heritage, Western governments and companies are ignoring this labour and the entitlements it creates. This position is summed up by Soleri et al. (1996, p. 24):

> [T]o those supporting farmers' intellectual property rights in their folk varieties, the effort and knowledge of cultural farmers involved in creating and maintaining folk varieties implies the need for recognition on an equal footing with that of plant breeders and molecular biologists. They see the communal effort in developing folk varieties as an integrated part of making a living over generations to be as legitimate as the individual efforts of scientists in formal, segregated work settings in the laboratory or field plot.

The belief that the work of generations of a community's farmers is as entitled to protection as the work of modern scientists is also voiced by Vandana Shiva (1990, p. 46):

> [P]atenting gives monopoly rights on life forms to those who manipulate genes with new technologies, totally disregarding the intellectual contributions of generations of Third World farmers, who for over 10,000

180

years have experimented in conserving, breeding and domesticating plant and animal genetic resources.

This argument has an intuitive plausibility, but it cannot be sustained. It raises several issues which can best be brought to the fore by a brief account of the labour theory of ownership.

The seminal account of the labour theory of ownership is, of course, that of Locke. In Locke's theory, the world's resources are *res nullius*: owned by no-one and available for appropriation by anyone. But when someone mixes their labour with a natural resource, they come to own that resource. This is because people own themselves[3] and by extension, their labour. They are therefore entitled to the product of that labour. In answer to the question of why mixing one's labour is a way of coming to own what you don't own, rather than losing something you do own (Nozick, 1974, pp. 174–5), Locke replies that labouring on something adds value.

The Lockean theory of ownership has been criticised on many grounds. For example, it has been argued that labour does not always create value; that the labour criterion is too vague; that the self-ownership thesis is atomistic, ethnocentric and fails to recognise that labour is a social activity; and that the labour theory cannot provide entitlements to *intellectual* property. We do not wish to enter those debates here; we simply want to see whether, even if the labour theory is sound, it can be consistently applied to the case of communities, IPRs and landraces. In a labour theory of intellectual property, it is the performance of a certain act – labouring – that creates an entitlement on the part of the labourer to certain rights concerning the product. The crucial questions here, then, are firstly, whether communities can be seen as actors capable of performing entitlement-creating acts and secondly, whether the development of landraces qualifies as such an act. We answer 'no' to both questions, which are dealt with here in turn.

a   Are communities capable of labouring?

There is a well-known tradition in political and social philosophy that sees communities, or nations, as independently-existing organic entities that have 'lives' of their own above and beyond the lives of their individual members. Clearly, communities do not have easily identifiable personalities like individuals, but some writers and traditions – from Rousseau to the romantics; from Hegel to the fascists – have pointed to a certain 'guiding consciousness', or the 'general will', in order to indicate that communities are greater than the sum of their individual parts. More recently, Van Dyke (1985) and Johnson (1991) have argued that communities often cannot be accounted for as an aggregate of individuals and that they have a moral standing of their own apart from the individuals that make them up.

However, even if we accept the controversial proposition that communities are independent entities that have independent moral standing, we do not have to accept that a community is capable of performing entitlement-creating acts – labouring – and thereby coming to own property. For entities to perform labour, they need to be in possession of many mental tools: a conception of the future; an idea of their own good and of how to achieve that good; the knowledge of how to perform the act that will achieve their aim; and so on. In short, to labour, an entity has to be capable of executing a rational plan aimed at some end. It is hard to see how a community, even if it were an independently-existing entity, could achieve such a thing without so much as a brain of its own. Unless we embrace mysticism, a community, then, just does not seem to be the kind of thing that can engage in labour. It is individuals and individuals only that engage in labour (whether by themselves or collectively) and, therefore, on the labour theory of ownership, only individuals can create entitlements for themselves and come to own property

However, let us for the moment embrace mysticism and assume that communities can perform acts of labour. Perhaps we can postulate that when an individual engages in labour, the community is in fact working *through* that individual in order to fulfil a community good that cannot be reduced to individual goods. What kind of IPRs would we have to award to respect the entitlements this community supposedly makes for itself when it engages in labour? If we view communities as singular entities and not as collections of individuals, the IPRs we grant would have to be permanent: the resources in question would be inalienable. The normal structure of an IPR is that an individual or collection of individuals is given a monopoly on an innovation for a limited period – in the 1994 Agreement on Trade-Related Intellectual Property Rights (TRIPs), the recommended patent length is 20 years. The claim under criticism here is that it is the ongoing labour of a community, over many generations, that creates an entitlement for that community in a landrace. If a modern-day IPR – a patent, or perhaps a plant variety right – was granted to a community in respect of a certain landrace, it would run out in, at most, 20 years' time; at which point any corporation would be able to use the landrace without permission to develop new, patentable, strains. Clearly, then, ordinary IPRs would be rather pointless and certainly would not meet the demands of campaigners like Shiva and Greaves and the communities on whose behalf they campaign. What they demand is an IPR that lasts for ever, permanently entitling the community in question to control over the landrace and payment for its use in plant breeding research and production.

There is another difficulty: if we consistently follow through this suspect organicist labour theory, the decision over who can use the germ plasm in plant breeding and biotechnological research and production becomes impossible to make, for the following reason. In order for communities to be able to allow 'their' landraces to be used, we would have to postulate that these communities, as singular organic entities, had a kind of Rousseauian 'general will' that could be discovered. But how do we discover this general will so that landraces can be transferred – with payment – from their original communities and used in modern plant breeding? Clearly, voting procedures will not suffice because a vote is merely the aggregate opinion of the individuals who happen to be eligible to vote at the time of voting, rather than the community's 'general will'. As the will of the community, assuming it exists, cannot be discovered by any practicable means, such as the vote of a single generation, any IPRs we grant over landraces would have to be, then, permanent, and the germ plasm itself would be forever inalienable: permission for outsiders to use landraces could never be granted. An absurd situation, unwanted by both plant breeders and traditional communities, but one we unavoidably arrive at if we take the community to be an organism.

The reader might object to our rejection of the idea that communities can create entitlements to property through labouring, on the grounds that there are many groups in society today that do own property as groups. The most obvious example is that of a joint stock company. There is no individual owner of such a company, yet the company, which is after all a group of people, certainly owns its capital. However, in this case it is clear that ownership can be reduced to individual ownership. Although the decisions in a joint stock company are taken by the directors and employees, ownership, and with it control in the final analysis, resides with the shareholders. The shareholders do not by any stretch of the imagination constitute an organic, collective entity; they are individuals who own a share of the company individually, because they have each bought equity in it. The shareholders' decisions at an AGM bind the company, but in a cultural community such snapshot decisions are not binding.

Perhaps a better example, and one that gets closer to the relevant issues, is an agricultural commune, where the wheat harvest is communally owned by the members and outsiders are not entitled to a share of it. Now we *could* account for this ownership by saying that the commune has laboured on the wheat, has brought it to harvest and is therefore sole owner of it. In this case it seems that the group has indeed created an entitlement for itself by virtue of its labour. However, we can more accurately characterise this situation by looking at it in individual terms. It is not that the commune, as a singular entity, has laboured on the wheat and therefore owns it qua commune. It is rather that all the individual members of the commune are entitled to a fair share of the wheat because it is they who jointly created it. The reason outsiders are not entitled to a share, on a labour-entitlement theory, is because they have not individually contributed.

One might ask why we cannot properly view landraces in the same way that we view the wheat in the above example: that is, why can't we say that landraces are owned collectively by the community, seen as a collection of individuals, rather than as an organic entity? We might justify this by arguing that the individual members of the community, who have been responsible for the creation of the landrace, have alienated their individually-acquired rights to their community, in the same way scientists employed by the plant breeding companies of the North alienate their rights over their creations to the company (in return for a steady wage and the materials with which to create new strains). However, in the case of both the commune and its wheat and the scientists who create new crop strains, the point is that a group of people come together at the same time to collaborate; to work towards a common goal (creating a product). Through that common labouring they come to own shares in the product. Under the labour theory, those who made the effort own the product of it – unless a previous contractual arrangement has been made whereby the product is signed over to another body, such as the company that employs the scientists who developed the new strain. Now, it is true that cultural community innovation is ongoing and individual farmers continually experiment with variations of the landraces they use, often in a systematic and rigorous fashion (Hobbelink, 1989, pp. 135–47; De Boef et al., 1993; Fowler, 1995, pp. 221–2). We are prepared to entertain the idea that this innovation could qualify as entitlement-creating. However, if it does, the entitlement would accrue to the individual farmer(s) and not the community. If the experimentation was collaborative, in an entitlement theory only the individuals who collaborated would be entitled to IPRs in the product; and in any case, the IPR would only apply to that particular new variation on the landrace – not the landrace as a whole. It might be possible for future landrace variations to be handed over, by the individual farmers who created them, to their communities, as long as these communities were given corporate status in law – in the same way scientists hand over their rights to the plant strains they create to their employers. However, such variations are not those used by Northern scientists in modern plant breeding. Campaigners want whole landraces, the traditional varieties used by cultural communities for centuries, covered by IPRs. In this case, the contractual arrangement necessary to hand over rights is clearly lacking: most landraces have been developed over generations by people who would never have heard of IPRs.

It might be said that this is unfair. There might be no conscious contractual arrangement, but in certain communities, the farmers of the past who were responsible for innovation in breeding regarded, unconsciously, the result of their work as belonging to their community: in which case, it could be argued that there was a *tacit* contract in operation whereby property rights in the new variation belonged to the community. In theory this gives us a way of attributing IPRs to communities in PGRs, on the basis of a labour theory. However, the conditions to be satisfied before such a theory could ground such IPRs would be so stringent that in practice it would have no effect. Firstly, the gradual innovations would have to be proved to be of sufficient novelty to count as entitlement-creating. Secondly, they would have to be proved to have been made in a sufficiently systematic and rigorous way. And thirdly, it would have to be proved that the community in question did indeed subscribe to a collectivist ethic whereby the fruits of individuals' labours were seen, *by those individuals themselves*, to belong to the community as a whole. Clearly, these conditions, though necessary if we want to ground IPRs in PGRs for communities in a labour-entitlement theory, are far too stringent ever to be passed in reality. The only meaningful way, then, to apply the labour theory of ownership to the landrace/community case is to regard the communities that created them as singular entities and not as collections of individuals. But, as we made clear in the above section (a), this leaves us with the unsatisfactory picture of organic communities preventing the rest of humanity from using their landraces for ever. It is important to note that we have by no means established here that cultural communities are not the rightful owners of landraces. We have merely shown that a labour theory of ownership is wholly inappropriate in accounting for this ownership. This is, at root, because the labour theory of ownership is a highly individualistic conception that, to say the least, does not comfortably apply to groups qua groups. However, there is another entitlement theory we have yet to consider. This is the proximity argument.

3   The proximity theory

There are two kinds of proximity theory we will criticise. The first simply asserts that proximity to germ plasm creates an IPR in it. Our argument against this version is equally simple – why should proximity entitle anyone to anything, never mind an IPR? The second theory attempts to answer this question by linking proximity to the harm principle. However, we argue that the renewable character of genetic resources means that such a principle cannot be applied in this case.

a   Simple proximity

This theory is a variation on the 'occupancy' theory of property. The argument is that local communities (and/or states) own their genetic resources by virtue of occupying the land on which they grow. Again, Shiva (1993b, p. 27) has implicitly expounded this argument by criticising American companies for exploiting wild genetic resources in the Third World:

...the US has engaged in unfair practices in the use of Third World genetic resources. It has freely spun millions of dollars in profits from Third World biological diversity, but has shared none of those profits with the original owners of the germplasm ... A wild tomato variety taken from Peru in 1962 has contributed $8 million a year to the American tomato processing industry, yet none of these profits have been shared with Peru.

The implication here is clearly that Peru – whether the state or its cultural people is unclear – ought, by right, to be paid a share of the profits from the sale of a tomato variety that was developed from a wild variety found in Peru. One advantage of this theory is that it appears to justify the intuition that Shiva and other writers have, that Third World states and/or communities (it is often unclear in which collectivity such writers are investing the rights and in fact the argument can be used to justify both state ownership and community ownership of germ plasm) own *wild* as well as landrace germ plasm. This is something the labour theory could never justify.

However, the mere fact that a gene, or species, exists in a particular place and is used by a set of people does not create an IPR over it. There may be an argument that the proximity of a people to a resource creates a property right attributable to that community in that resource. For example, many people would intuitively think that the Ogone people of Nigeria are the rightful owners of the natural oil resources that lie beneath the Ogoneland region of Nigeria and that these property rights are being violated by the Nigerian government who are currently claiming the oil for themselves. We are prepared to accept that some people find it perfectly plausible that mere proximity to a resource creates a moral entitlement to it. However, we find it implausible that mere proximity to, or even use of, a certain type of plant by a people creates an *intellectual* property entitlement in that *type* of plant. To pick on proximity as a qualifying principle goes against the whole sense of entitlement theories of all kinds – that there is something morally compelling about the relationship of the claimant to the resource in question. Proximity is quite clearly an accident and therefore morally irrelevant. While proximity might be a useful principle in distributing property under a consequentialist theory, its arbitrariness precludes it from providing compelling moral entitlements to intellectual property.

b   Proximity-harm

Perhaps it could be argued from the proximity principle that natural resources could be considered the legitimate property of local people in that local people would suffer in some way if these resources were taken away from them. On this argument the harm principle could be used to ground a property right. This involves treating property rights as secondary to another, primary, moral principle (harm) but still links property rights to proximity; proximity is what would entitle particular communities to particular resources. If a forest provides the essentials of life for a people who inhabit it, for example, it might be considered their property. This fact would prevent outsiders' destroying or degrading the forest by cutting down its trees for timber and thereby destroying the community by taking away its livelihood.

In the case of genetic resources, however, we cannot use the harm principle in this way. It seems clear that cultural communities do not suffer in any way from the mere removal of genetic resources from their land. This is because genetic resources, if handled correctly, are *renewable*; samples can be taken away and modified by biotechnologists without the inhabitants of the host country being harmed at all (Kloppenburg, 1988, pp. 214, 222, and 222). Suman Sahai argues that (1994, p. 87):

> [I]f the copper found in Canada is Canadian and the coal found in Germany is German, then by the same principle, genetic resources found in the third world will have to be considered their property. There cannot be two sets of rules.

If Sahai means that the Third World is entitled to property rights in particular manifestations of genetic resources (i.e., actual plants) then he may have a case, but if he means that the Third World is entitled to *intellectual* property rights in these resources, on the same set of principles that Canada is entitled to its copper, he is wrong. The two circumstances clearly differ markedly; the principle that a state is entitled to ordinary property rights in the physical resources found within its territory does not entail the quite separate principle that states/peoples are entitled to intellectual property rights in the living resources found within its borders.

Similarly, Shiva (1991a, p. 2746) complains that US firms have profited to the tune of $66 million dollars by developing varieties out of Third World germ plasm, which they acquired for nothing. She concludes that the countries in which this germ plasm is found are thereby entitled to a share of the profits. But the fact that US firms have made a $66 million dollar profit on germ plasm does not mean 'donor' countries/cultural communities have lost $66 million dollars; and just because germ plasm from traditional varieties and wild species is now being used in crop strains worldwide does not necessarily mean that cultural communities have to stop using the original strains or living their traditional lifestyles.

However, there is a more compelling argument behind the proximity-harm principle. Commentators argue that certain groups might be harmed indirectly when PGRs taken from them are used in plant breeding and food production. This is because they might displace, with varying degrees of directness, the original crops and their growers. For example, one of the biggest worries of campaigners involved in PGRs is that new techniques of vanilla production based on tissue culture will obviate the need for growing the actual vanilla plant; it will be possible to produce the required essence in vats in the industrialised world. This will spell ruin for thousands of vanilla growers in the developing world, particularly in Madagascar. Furthermore, the germ plasm used in this new method of production will originally have been taken from the farmers that it will probably bankrupt. It seems that it would be only fair for some sort of compensation to be paid and that attributing IPRs in the original germ plasm to the farmers whose livelihoods have been lost would be a reasonable way to go about this.

There is, however, an imbalance between the principle that a loss in utility ought to be compensated and the proposed solution. Firstly, not all acquisitions of PGRs by breeders will harm the original holders. In the case of vanilla, the harm will be relatively direct and indisputable: vanilla essence produced using tissue culture will, in all likelihood, displace 'real' vanilla from the market. In the case of landrace crops used in the breeding of modern varieties, however, the harm caused may be extremely difficult to unravel. New varieties undoubtedly displace traditional varieties, but do not necessarily damage the farmers who grow them. It may be that the uniformity of modern varieties and the level of inputs – fertilisers, pesticides and herbicides – they demand has a tendency to push out peasant farming and promote the growth of 'agribusiness', but this is difficult to prove and would be resisted by the companies who produce modern varieties. Indeed, they might argue that farmers actually benefit from the superior qualities of their products[4]; they can state with some degree of plausibility that it is farmers themselves who choose to use these new strains and that if they were harmed by them they wouldn't choose to use them (Murphy, 1988, p. 214). Certainly, plant breeding companies would be unlikely to accept an IPR regime based on the idea that their products are harmful.

Secondly, unlike with non-renewables such as copper and coal, it is not only the acquisition of PGRs, but their use and marketing, coupled with certain economic conditions, that creates the harm. This means that it would not make sense to offer payments of compensation at the point of acquisition: they would have to be made retrospectively, only after the harm had been caused. This has the effect of delinking compensation from proximity and making clear that, at root, the proximity-harm principle is a utilitarian principle. For the application of the principle to be fair and impartial, compensation would be due to all those who are harmed by the acquisition and use of a particular set of PGRs. In the case of vanilla and tissue culture mentioned above, to follow through the principle in full would mean compensating all vanilla farmers around the world when the new methods of production displace them from the market, i.e., when the utility is lost. Yet attributing IPRs to the communities or states from where the original germplasm came would mean compensating only some of those harmed: thousands of vanilla growers would go uncompensated simply because they were not part of the community or state deemed to have IPRs in the original germ plasm. Using the harm principle entails compensating all those who lose utility; IPRs would fail to ensure this. Harm might, then, be a useful principle in justifying and organising compensation for those harmed by the development of biotechnology, but cannot coherently or fairly be used to justify giving IPRs in original germ plasm.

4   The ownership of folk knowledge

Claims are also made by some writers that cultural communities are entitled to IPRs in their traditional knowledge; that is, their skill and expertise relating to the various properties of living resources and how to harness them for human benefit. Some writers have complained that indigenous knowledge is being plundered by Western transnational corporations who are then patenting this knowledge for themselves. Holla-Bhar and Shiva (1993, p. 224) present a convincing account of how the US corporation W.R. Grace has hijacked cultural knowledge of the properties of the neem tree and now holds patents based very strongly upon this knowledge. Similarly, Brush (1993, p. 659) argues that traditional knowledge is useful and is therefore entitled to IP protection on the same basis as Western scientific knowledge:

> [I]f commoditizing Western scientific knowledge is justified by the wider public interest served, then cultural knowledge is likewise entitled to protection as intellectual property because it is useful ....

However, Brush is illegitimately mixing utilitarian and entitlement theories of intellectual property. He is right to point out that the prime justification for IPRs in the West, whether it is valid or not, is that they increase the rate of innovation and thereby increase overall social utility. But the point is not that Western scientific knowledge is 'useful' and therefore somehow 'entitled' to protection; rather that *legally protecting* Western scientific knowledge is useful overall. The mere fact, then, that cultural community knowledge is useful does not, either on an entitlement or utility theory, mean that it ought to be legally protected. An entitlement theory finds the consequent rise or fall in utility irrelevant, while utilitarian theory has no use for the concept of entitlement. For the utilitarian argument to be valid, it would have to be shown not only that cultural community knowledge is useful, but that protecting cultural community knowledge through IPRs would increase overall utility.

Under the labour-entitlement theory of intellectual property, which is the theory implicitly being used by Holla-Bhar and Shiva and other supporters of IPRs in traditional knowledge (Greaves, 1995), it is a single act of creation that creates an entitlement on the part of the creator to the exclusive benefit of the fruits of the creation. With traditional knowledge, however, there is no single act of creation: traditional knowledge is not the discovery of a single person or group of people, but is the result of centuries of collective experience – in which case there was never any one person or group of persons *entitled* to private property in this knowledge. Moreover, even if there was an individual act of creation, there can be no present entitlement to IPRs in traditional knowledge. Intellectual property is intended to give inventors and creators a monopoly on the fruits of their inventions and creations *for a limited time*. The time limit is not an arbitrary legal notion, but, if we adhere to the entitlement theory, is crucial, because it is recognised that invention is a partly social phenomenon: inventors rely on education, a political and social environment conducive to innovation, adequate materials and so on, which are all provided socially (Spencer, 1970, pp. 140–2). It is also likely that others would have come up with the same invention sooner or later (Nozick, 1974, p. 182). For this reason, ideas and inventions turn into common property after a certain duration. In the light of these facts, it seems clear that traditional knowledge, such as the Indians have in the properties of neem, does not qualify as intellectual property on an entitlement theory. Even if traditional knowledge was, originally, the product of a single person or group, it seems that the time limit would have run out by now and the knowledge would have become common. It is certain, furthermore, that the individual creators/ discoverers of such knowledge, if they existed at all, would now be dead. As IPRs are not inheritable,[5] an entitlement theory cannot ground property rights for the present members of cultural communities in the knowledge of their communities.

Folk knowledge, then, is not intellectual property under an entitlement theory. Thus, people are mistaken when they complain of Elvis Presley 'stealing' black people's music and gaining a fortune from it;[6] the people of Vanuatu are wrong to claim royalties from bungee jumpers (they are claiming royalties on the grounds that bungee jumping is based on the traditional Vanuatan practice of vine-leaping (*Daily Telegraph*, 1995, p. 5)); and Shiva (1993a, p. 26) (along with many others) is wrong to say that Western corporations have 'robbed the Third World of its ... biological knowledge'. Knowledge is not a physical thing; cultural communities still have their knowledge no matter how many others have used it. It cannot be stolen. This does not, of course, mean that W.R. Grace's neem-related patents are necessarily just; if they were simply modern scientific formulations of already-existing traditional knowledge, they are faulty and should never have been granted. Whether discoveries of genes and substances ought to be patentable is also open to question. But traditional knowledge cannot count as intellectual property on any entitlement theory and no wrong is perpetrated by a company merely producing products based on traditional knowledge.

5   Conclusion

We have not conclusively shown that we ought not to grant IPRs over genetic resources to cultural or local communities; merely that such communities do not hold *moral entitlements* to such property. There may be other, consequentialist, reasons for us to grant IPRs: we may think they are necessary to preserve landraces, or to preserve traditional ways of life, or to promote sustainable agriculture, or to prevent exploitation, or to increase the rate of innovation. Attributing IPRs to communities, given corporate status as a legal fiction, might be an effective way to go about achieving these aims, but this is another question.

We have shown that labour-entitlement theories are only applicable to communities if we either subscribe to suspect organicist ideas of communities as independently-existing entities (ideas that, even if acceptable, preclude the possibility of communities ever being able to allow the outside use of the PGRs they supposedly own), or if we rely on a number of conditions that would be difficult, perhaps impossible, to satisfy in reality. As for proximity, we regard this as accidental and therefore antithetical to the general thrust of an entitlement theory: that there ought to be a morally compelling relationship between claimant and resource. If we add harm to proximity, we gain a more morally satisfying principle, but one that does not lead to the proposed solution. Finally, we have argued that the 'traditionality' of traditional knowledge – the fact that it is common knowledge, the product of collective experience without a single act of creation – precludes its being seen, from the point of view of an entitlement theory, as intellectual property. It seems, then, that entitlement theories just cannot be meaningfully and consistently applied to communities. An entitlement theory of IP demands two things: an individual creator or group of creators and an identifiable creative act. Neither of these requirements are satisfied in the case of communities and PGRs and their associated knowledge.

Notes

1   The Kani agreement suggests that this practice is, in fact, on its way out. It was, however, the dominant way of obtaining PGRs from gene-rich countries for many years.
2   Landraces are varieties of domesticated crops that have been created through centuries, even millennia, of artificial selection by farmers. They are also called 'traditional' or 'folk' varieties.
3   Strictly speaking, according to Locke, people are 'owned' by God, but a secular version of the Lockean theory entails self-ownership.

4   Murphy (1988, p. 214) points out that most elite, high-yielding, varieties are adapted to the country they are grown in.

5   At present the only kind of IPRs that could conceivably be said to be inheritable are copyrights, which tend to run for the author's life plus a fixed number of years, depending on the country. The posthumous royalties are due to the author's estate, which usually means the family (Hughes, 1988, p. 323). Patents and other IPRs are most certainly not inheritable; to make them so would be wholly novel.

6   If Elvis used other people's songs, or tunes, it would be a different story; to our knowledge, he didn't do this.

References

Brush, S. (1993), 'Cultural Knowledge of Biological Resources and Intellectual Property Rights: The Role of Anthropology', *American Anthropologist*, Vol. 95, No. 3, pp. 653–71.

Brush, S. (1996), 'A Non-Market Approach to Protecting Biological Resources' in T. Greaves (ed.), *Intellectual Property Rights for Cultural Peoples: A Sourcebook*, Society for Applied Anthropology: Oklahoma City.

Crucible Group (1994), *People, Plants and Patents: The Impact of Intellectual Property on Biodiversity, Conservation, Trade and Rural Society*, International Development Research Centre: Ottowa.

Da Carta E Silva, E. (1995), 'The Protection of Intellectual Property for Local and Cultural Communities' *European Intellectual Property Review*, No. 11, pp. 546–9.

Daily Telegraph (1995), 'Island seeks bungee profits to help the economy bounce back', 16 Nov, p. 5.

De Boef, W., Amanor, K., Wellard, K. and Bebbington, A. (eds) (1993), *Cultivating Knowledge: Genetic Diversity, Farmer Experimentation and Crop Research*, Intermediate Technology Publications: London.

Food and Agriculture Organisation (1989), *Report of the Commission on Plant Genetic Resources*, FAO: Rome.

Fowler, C. (1995), 'Biotechnology, Patents, and the Third World' in V. Shiva and I. Moser, (eds) *Biopolitics*, Zed Books: London.

Greaves, T. (1995), 'The Intellectual Property of Sovereign Tribes', *Science Communication*, Vol. 17, No. 2, pp. 201–13.

Greaves, T. (ed.) (1996), *Intellectual Property Rights for Cultural Peoples: A Sourcebook*, Society for Applied Anthropology: Oklahoma City.

Hobbelink, H. (1989), *Biotechnology and the Future of World Agriculture*, Zed Books: London.

Holla-Bhar, R. and Shiva, V. (1993), 'Intellectual Piracy and the Neem Tree', *The Ecologist*, Vol. 23, No. 6, pp. 223–7.

Hughes, J. (1988), 'The Philosophy of Intellectual Property', *Georgetown Law Journal*, Vol. 77, pp. 287–366.

Jayaraman, K.S. (1996), '"Indian ginseng" brings royalties for tribe', *Nature*, Vol. 381, p. 182.

Johnson, L. (1991), *A Morally Deep World*, Cambridge University Press: Cambridge.

Kloppenburg, J.R. (ed.) (1988), *Seeds and Sovereignty: The Use and Control of Plant Genetic Resources*, Duke University Press: Durham, North Carolina.

Margulies, R.L. (1993), 'Protecting Biodiversity: Recognising Intellectual Property Rights in Plant Genetic Resources', *Michigan Journal of International Law*, Vol. 14, No. 2, pp. 322–56.

Nozick, R. (1974), *Anarchy, State and Utopia*, Blackwell: Oxford.

Shiva, V. (1990), 'Biodiversity, Biotechnology and Profit: the need for a People's Plan to protect biological diversity'. *The Ecologist*, Vol. 20, No. 2, pp. 44–7.

Shiva, V. (1991), 'Biotechnological Development and Conservation of Diversity', *Economic and Political Weekly*, 30 November, pp. 2741–5.

Shiva, V. (1993a), 'Biodiversity and Intellectual Property Rights', *Earth Island Journal*, Vol. 8, No. 4, pp. 25–7.

Shiva, V. (1993b), *Monocultures of the Mind: Biodiversity, Biotechnology and the Third World*, Third World Network: Penang.

Soleri, D. and Cleveland, D. with Eriacho, D., Bowannie, F., Laahty, A. and Zuni Community Members (1996), 'Gifts from the Creator: Intellectual Property Rights and Folk Crop Varieties' in T. Greaves (ed.), op. cit.

Spencer, H. (1970), *Social Statics*, Gregg International: Farnborough.

United Nations Environment Programme (1992), *Convention on Biological Diversity*.

Van Dyke, V. (1985), *Human Rights, Ethnicity and Discrimination*, Greenwood Press: London.

12 The merchandising of biodiversity

Joan Martinez-Alier

1 Introduction

Indigenous groups have accumulated an enormous body of knowledge about biological diversity and peasant farmers have been selecting and improving seeds for a long time. This knowledge of natural biological diversity and the conservation of agricultural diversity in situ has seldom been valued in economic terms. According to most economists, the fact that genetic resources have not been appropriated and treated as merchandise is the reason for the 'genetic erosion' that is taking place today; things without an owner or a price are treated as if they are worthless. It is now being proposed that access to natural genetic resources should have an economic price and that 'farmers' rights' should be recognized so that farmers' conservation work can be rewarded. In opposition, a growing popular ecological movement seeks to defend agricultural biodiversity, not through the market (where the poor are weak and future generations are not represented), but through political and social movements favouring ecological agriculture. This article weighs the efficacy of these two approaches in preserving the biological diversity upon which we all depend.

2 Genetic erosion

Modern agriculture, based on varieties that have been improved by non-traditional techniques, greater production per acre and high fossil fuel energy input, has led to the reduction of biological diversity in agricultural systems.[1] According to the FAO (1993), 75 per cent of plant genetic resources have disappeared over the last decades. A study by Renée Vellvé (1992) shows that modern agriculture in Europe replaces diversity with uniformity and security with vulnerability, leading to biological impoverishment.

One example: the development of hybrid corn 50 years ago, and its spread in the United States, led to biological impoverishment and required a continuous input of genetic material from areas where these new and uniform varieties are still not cultivated. There has been no research into free pollination varieties, which over time would have permitted yields as high as those of hybrid corn in the US. The farmer, however, would have controlled the seed.[2] The economics of technological change took research and development of hybrid corn as one of its classic examples (Grilliches, 1958). The complementary inputs for this monoculture were accounted for on the basis of their market prices alone; negative economic externalities from fossil fuels used, increased soil erosion and loss were ignored. The development of hybrid corn, varieties of wheat and rice, set off the current process of genetic erosion within an agricultural system based on mechanization and infield monoculture.

In Western agricultural economics, it is difficult to find any (or even any attempt to construct indices of) loss of agricultural biodiversity. The introduction of new varieties has been seen as technical progress, the monetary costs of which are more than compensated for by greater production, due to greater agricultural input. Meanwhile, genetic resources are increasingly falling into the hands of multinational industrial companies. The efforts of public institutions to store these resources ex situ in gene banks have run into many problems. The alarm over genetic erosion has grown, resulting in new policies for ex situ and in situ conservation.

What are the reasons for genetic erosion in agriculture? Is the expansion of the market the main culprit? Or would the extension of the market be a solution? Some argue that an ecologically extended market can incorporate ecological costs into market prices, in effect denying that conflicts between economic-chrematistic reasoning and ecological reasoning need to be solved politically. Their position is that these conflicts can be solved by ensuring that the products of ecological agriculture obtain higher prices in new, special-ized markets. Others argue that the important question is, who can best voice the conflict between ecological reasoning and economic-chrematistic reasoning? Put another way, can the movement for ecological agriculture become a political ideology capable of mobilizing peasants skilled in the sustainable use of biological resources, soil, water and current solar energy?

3   Struggles to control seeds

The GATT negotiations tried to impose acceptance of intellectual property rights to commercial agricultural seeds upon India, while the NAFTA agreement between Mexico and the United States could still be the final blow to traditional agriculture in southern Mexico. Fortunately, there has been another party involved, not only in those countries of the South with the greatest biological diversity, but also in Europe. The most important work of conservation has been carried out — whether by farmers, individuals or local groups — as popular, ecological initiatives, however inadequately financed and with little social recognition.

Biological diversity was one of the most important questions at the Rio Conference, but only now are people in the poor countries becoming aware of its value to the rest of the world. Some of these poor countries include Vavilov's centres of origin for agricultural diversity. In these countries, there are still poor farmers who are experts in the traditional selection and improvement of plants and who practice agriculture with few external inputs, based on hundreds of local varieties (Cooper, Vellvé and Hobbelink, 1992; Querol, 1987 and 1992). The threat to this agricultural diversity comes mainly from the extension of the market and from the fact that decisions relating to production are increasingly made on the basis of priorities indicated by prices. If profits increase with the introduction of modern agricultural techniques and the so-called high-yield varieties, the varieties that have been improved by traditional methods can be preserved only through political action. With the support of the FAO, a new consensus is forming in the South that the work by traditional farmers in the creation and conservation of agricultural genetic resources should be recognized and compensated. Poor people's ethnobotanical knowledge of diversity has been seen as a central component of an ecological agriculture based on permanently developing indigenous and peasant knowledge (Richards, 1984; Guha and Gadgil, 1992; Toledo, 1988, 1989 and 1991; Posey, 1985; Descola, 1988; Rocheleau, 1991).

Pride in traditional ecological agriculture, an excellent model of 'clean' technology, is growing, at a time when there is much discussion of technology transfer. This pride is accompanied by awareness of the fact that little was paid for traditionally improved varieties. A low price was paid in the peasant markets where the seed was collected for ex situ gene banks and nothing at all was paid for the collection of genetic material under cultivation. Nor was anything paid for the medicinal plants discovered and cared for by indigenous cultures, then used, developed and charged for by pharmaceutical companies. Royalties are even charged for medicines protected by trademarks and patents. The appropriation of genetic resources ('wild' or agricultural) without adequate payment or recognition of peasant or indigenous knowledge and ownership over them has been described as 'biopiracy'. The name (given by Pat Money around 1993) is increasingly being used and it reflects the feeling of injustice. The fact is not new, the feeling perhaps *is*.

Unlike medicines, modern improved commercial seeds so far have not been patented. Protection against their duplication has been achieved by the UPOV[3] system and by the sale of hybrid seeds that do not breed true. But the new legal framework demanded by the biotechnology industry will allow the patenting of 'life-forms', including agricultural genetic resources. This is the reason why GATT is imposing the international recognition of patents (or their equivalents) on 'new' genetic resources, in the same way that it has always tried to enforce the recognition of patents on medicines.

Now, under the Biodiversity Convention (Rio, 1992) there is much talk of paying appropriately for genetic resources, not only because of equity considerations but also in order to give an incentive for conservation. The question remains whether the commercialisation of genetic resources will benefit conservation. Payments are very low, given the social asymmetry between the multinational companies which engage in 'bioprospecting' and the indigenous groups – see for instance the case study of the transactions between Shaman Pharmaceuticals and a group of Pastaza Quichuas in Ecuador (Reyes, 1996). Such low monetary incentives will not offer a more competitive advantage to traditional uses of the forest than do the new industries: wood extraction, mineral exploration and extraction and ranching. Moreover, the 'bioprospecting' firms, even if they are used to the relatively long periods of maturation of investment of the pharmaceutical industry, have short time horizons – of 40 or 50 years at most – compared to the very long periods of tens of thousands of years relevant for biological evolution.

At present, the FAO International Undertaking on farmers' rights and the international Treaty on Biodiversity signed in Rio will probably produce a very small international fund for in situ conservation of plant genetic resources.[4] International policies on biodiversity will be guided by the theory of an 'optimal portfolio' – how to conserve the 'appropriate' amount of biodiversity, neither too little nor too much (because it would be chrematistically expensive).

The main reason for in situ conservation is the potential for coevolution of plant genetic resources. But how many traditional agriculturalists, with their complex agroecosystems of still unknown genetic wealth, should be preserved and adequately compensated? One billion? Half a billion? After opening up the box of agricultural biodiversity, pro-peasant NGOs will not be intellectually and politically pacified by an international fund of $50 or $100 million per year (about 10 cents per farmer), the amount that is supposed to actually implement 'farmers' rights'! It would be a pity if all the struggles for agricultural biodiversity and 'farmers' rights' yielded at the end a small international fund, managed by the World Bank and the CGIAR, consisting of a few traditional farming in situ conservation areas in the world.

There are also historical aspects of 'farmers' rights', connected to the increasing discussion of the 'ecological debt' of the North to the South. Let us imagine the case of a group of people outside the generalized market system, with ancient and proven healing methods, part of their vast repertory of medicinal knowledge, both botanical and zoological. This indigenous knowledge is not built up in a single generation, nor is it static and unchanging; there is always experimentation and improvement.[5] Let us now suppose that this knowledge, along with the relevant materials, is transferred to an outside group, without anything being given in exchange. This could occur by means of scientific research, by the work of missionaries of another religion or by simple political and economic exploitation (public or private). Now let us suppose that this other group translates and absorbs this knowledge into its own culture, manipulating the materials so that its members can apply them in its own system of medicine. Furthermore, suppose that through direct political imposition or generalized incorporation into the market, the group responsible for this exploitation manages to ensure that the primitive tribe must pay hard cash for the re-elaborated curative materials and medicinal knowledge. This is what has been happening in the medical and pharmaceutical industries. We can accept the superiority of modern medicine while at the same time accepting the description above and understand that something similar is now taking place in the case of agricultural seeds, due in part to the GATT negotiations which include TRIPS (Trade-Related Intellectual Property).

In India, the KRRS Association (Karnataka Rajya Raitha Sangha), in cooperation with the Third World Network, coordinated by Vandana Shiva, organized a variety of massive actions against transnational seed companies in 1993 and 1994. The protest was against the possibility that the Indian state might establish powerful systems of intellectual property rights to 'improved' seeds as a consequence of the GATT negotiations. In this case, the farmers would no longer be able to produce these seeds and exchange them among themselves. Furthermore, they have never received anything in exchange for their work of conserving and improving their own seeds over many generations. One response was the destruction of Cargill Seeds' installations in Sirivara, Bellary District, Karnataka. There was also strong opposition to W.R. Grace & Co's project to set up an installation to manufacture biological pesticides based on the seeds of the *Nim* tree *(Azadirachta indica)*, long used as an insecticide. The question is thus forcibly raised, who does biological diversity and indigenous agricultural knowledge belong to? Can it properly be acquired without payment by the North and then returned in the form of patented seeds and pesticides? Even if a pesticide with the properties of *nim* seeds is chemically synthesized, making it unnecessary to gather them in India, has this traditional Indian knowledge no value at all?

In India, there are tens of thousands of varieties of rice, many in danger of being lost; other varieties were collected without payment by 'gene banks' (banks that neither pay interest nor return capital to depositors), especially the Philippine International Rice Research Institute, where the rice varieties used in the 'green revolution' originated. These plant collections (like others held at CGIAR-affiliated CIAT in Colombia, CIP in Peru and CIMMYT in Mexico) are now at risk or being patented to the benefit of international seed companies.[6]

The emerging ecological movement related to agricultural diversity raises two questions. The first is the recognition of (and payment for) farmers' rights for the genetic resources that they have conserved and improved in situ. The second is that of favourable, if not free, access to the varieties that have been conserved and improved ex situ, based on the reasoning that the precursor genetic materials originated within traditional agriculture and that no one paid anything for them. At the same time, the governments of the South are waking up; the precursor genetic resources were, until recently, 'world heritage'. But now several states are moving quickly to declare them state property on the basis of their interpretation of the Rio de Janeiro Biodiversity Treaty. But it is doubtful that ownership by the state will actually favour poor farmers or indigenous communities.

4   Agricultural biodiversity as 'cultivated natural capital'?

Agricultural diversity cannot be understood unless we consider the entire human ecological complex of the societies that have created, conserved and raised this wealth of genetic resources. There is no question that these resources are of great value, although it is difficult to assign a monetary value to them. The question is whether genetic resources in general (wild resources, improved traditional varieties, modern varieties and genetically engineered varieties) should be commercialized or whether they should continue to be 'world heritage'. So far, the resources produced by traditional selection and improvement of plants and then collected from cultivation have not been paid for, whereas the companies that sell modern improved seeds insist on being paid. The products of genetic engineering will not only be sold; they will also be monopoly products as a result of a patent system.

Although the Biodiversity Treaty signed in Rio[7] recognizes that peasants and indigenous peoples have used and conserved these genetic resources since time immemorial, the treaty does not ensure their ownership and management rights to these resources. The treaty also failed to include a critical part of the planet's biological diversity; that held by national and international gene banks. This was due to pressure exerted by the US at the preparatory meeting in Nairobi on 22 May, 1992. The inclusion of germ plasm held by gene banks within the scope of the biological diversity treaty would have forced the signatory industrialized countries to share the profits made from these seeds or germ plasm with the poor countries, thus attacking the commercial interests of the big seed companies. Modern so-called improved varieties (as if traditional varieties had not been improved since the neolithic period) cannot function without a continuous supply of new genetic resources to confront new pests and new environmental conditions. But modern agricultural varieties are chrematistically more profitable. Thus, the increase

199

in production for the market spoils the very conditions necessary for this production, namely, agricultural biodiversity. As a result, a new ecological movement has arisen to fight against this degradation of the environment.

Among the economists who are more open to ecological matters, the value of wild biological diversity has been considered in the following terms. Its conservation has an immediate use, as it is a genetic resource for the chemical and pharmaceutical industries, as well as for the seed companies. There is also a possible future usefulness, called an 'option value'. Finally, there is an 'existence value' shown, for example, in the membership fees that members of Greenpeace pay to save the whales. Members want to save the whales, not so that people can kill them to extract the oil and flesh, nor to conserve a stock for future exploitation, but rather because they are endangered species with a right to survive.[8]

The main aim of organizations like the World Wildlife Fund has been to conserve natural biological diversity, which receives greater attention than agricultural or agroforestry diversity in the IUCN conservation strategy (McNeely et al, 1990). But there is a complementary relationship between wild and agricultural biological diversity. Agricultural genetic resources could be called 'cultivated natural capital' and cannot fully be substituted by the capital goods (including improved seeds) used in modern agriculture. This 'cultivated natural capital' needs to be complemented by natural capital, i.e., the 'wild relatives' of cultivated plants.[9] Some ecological economists call natural resources 'natural capital'. This change in terminology (from natural resource to 'natural capital') indicates the absence of depreciation or amortization for natural resources that would be obvious if they were called natural capital. It also indicates the problematic nature of the substitution of natural resources by capital.

Nevertheless, I am not in favour of using the term 'natural capital'. Among the natural resources of orthodox economics there are resources that are neither merchandise nor produced like merchandise (e.g., the genetic resources of traditional agriculture and wild biological diversity). Others, such as land, are not produced as merchandise but are sold or rented as such. Assigning the name 'natural capital' to all natural resources may betray the intention to consider them all as merchandise. In other words, is talking about 'natural capital' the same as talking of 'nature as capital?'

The market (or surrogate markets) cannot give convincing values to future events, which are uncertain and irreversible. Sometimes it is argued that a negative externality has a value equal to the cost of repairing the damage. For example, the price of chemical contamination (which the market does not value) would be the cost of restoring the contaminated site to its former condition. If we try to value the loss of biological diversity using this criterion we come up against the problem that the loss is irreversible. The criterion could be modified as follows: the price of biological diversity is what it would cost to maintain it, not only in terms of the costs actually incurred, but also in payment for tasks that until now have been unpaid and in terms of opportunity costs and benefits (e.g., the cost of a lower level of agricultural production or the cost of not destroying the rainforest, which will also have beneficial effects on the climate). This is not the same as creating legal rights to biological diversity and then organizing a market in these rights. Nor is it the same as a cost-benefit analysis in present-value terms of the conservation of biodiversity.

5   Biodiversity and marketplace realities

Activists in favour of ecological agriculture are against the patenting of 'life form'. They are in agreement with many other ecologists who fear that the development of biotechnology, with all its promises and threats, will only be driven by the logic of the marketplace. Those in favour of ecological agriculture believe that the CGIAR's Agricultural Research Centres should not patent their genetic resources. They are against intellectual property rights, because they do not believe that this is the way to defend and reward agricultural diversity. They do not even all agree with the payment of 'farmers' rights', which are not the equivalent of buying intellectual property rights, but more like fees or prizes for professional services. While patents, copyrights, trademarks and intellectual property rights in general are the property of their creators and inventors, prizes and honours are other ways to stimulate creativity, reward time and money invested and to compensate for inventions.[10] 'farmers' rights' belong in this category, and would serve to give the necessary incentive to ensure the conservation and development of agricultural diversity. Rather than paying 'royalties' for traditional seeds, it would be better to consider *all* genetic resources as 'world heritage'. At the same time, we need to introduce social and legal obstacles to dangerous or absurd applications of biotechnology (like increasing resistance of plants to pesticides instead of pests) and to establish economic compensation by means of product prices (or income transfers) for ecological agricultural producers using 'clean' technologies and few inputs, to give them incentives to maintain and develop traditional diversity.

Should genetic resources become merchandise so that an ecologically extended marketplace will conserve them? Here I shall insist that it is impossible for future generations to participate in the present market (see Martinez-Alier, 1990). In addition, I would like to point out that market values depend to some extent on the current distribution of power and income. Who would receive these 'farmers' rights' if they were sold in the market? The farmers' organizations? The individual farmers? The governments? What price would be put on them? The truth is that peasants and indigenous groups would sell their hypothetical 'farmers' right' cheaply, not because they have (until now?) attributed a low value to their work and agricultural knowledge, nor because they give little value now to the benefits of biological diversity for future generations, but because they are poor. If the poor sell cheaply, then there is no reason to trust that prices in an ecologically extended market will be an effective instrument of environmental policy. There is obviously a need for environmental policies based on social movements, going beyond an ecologically extended market.

An example of this is the agreement that the Instituto Nacional de Biodiversidad (National Biological Diversity Institute) (INBio) of Costa Rica and the Merck company reached in 1991 (Gamez, 1992; Reid et al., 1993). In this case, what is sold are not agricultural genetic resources, but wild ones; but the case is relevant to my argument. What INBio is selling is a service, the collection and preparation of a large number of samples of biological diversity, samples of the plants, insects and microorganisms that the Institute has access to in the conservation areas of Costa Rica. INBio (which disguises itself as a private organization, but acts under state protection) has free access to these resources and only pays the cost of collection by 'parataxonomists' (who possess their own knowledge, which they sell cheaply) and the cost of preparing the samples. INBio does not pay the direct costs of establishing and guarding the natural parks nor the opportunity cost of maintaining these wildlife reserves. The World Resources Institute praised this 'recent agreement between a major pharmaceutical company and Costa Rica which deserves to be widely copied' (World Resources Institute, 1992, p. 10).

However, the agreement caused uneasiness in Latin America because, inter alia, Costa Rica shares many of these genetic resources with neighbouring countries. The agreement implies the recognition of rights to genetic resources ('wild' ones, in this case), but it does not guarantee that traditional wisdom and conservation of biological diversity, as such, can compete with other forms of land use that are more profitable in the marketplace. Under the terms of the agreement, Merck will pay around one million dollars over two years for rights to chemical screening of a large number of samples prepared by INBio from a large area of Costa Rica that is protected. In addition, Merck will pay royalties on profits from any commercial products. Without further costly conservation measures, such as legal regulation and police vigilance (to be paid for by the Costa Rican authorities or other bodies) to complement the local population's interest in conservation, the limited chrematistic incentive provided by Merck would be insufficient to prevent deforestation and genetic erosion. Merck is a commercial company, with a relatively short-term outlook, extending at most to a few decades. Furthermore, it is to be expected that Costa Rica will sell cheaply. Why has Costa Rica, the classic banana republic, sold bananas cheaply to United Fruit, Standard Fruit or Del Monte? Because it wanted to? Of course not! If Costa Rica cannot get a good price for bananas, how can it get a good price for biological diversity? The poor sell cheaply. And future human generations – and other species – cannot even come to the market.

6   The defence of ecological agriculture outside the market

The environmental effects of modern agriculture (loss of genetic resources, destruction of non-renewable energy from fossil fuels) make it doubtful that modern agriculture is really more productive. Increases in productivity (per acre or per hour of work) are measured by subtracting the inputs from what is produced and then dividing the result by the quantity of the input whose productivity is being measured. But the prices of production and inputs are badly measured because they do not include externalities, including the destruction of the conditions necessary for agricultural production.

At this point, there are two possible paths. The first, easier to set out on, but which may soon become extremely narrow, tries to reconcile economic-chrematistic reasoning with ecological reasoning. For example, by means of a Green Label, the products of ecological agriculture may obtain higher prices, as long as there is a demand for these differentiated products. Victor Toledo (1992) has expressed this idea as follows:

> [E]cological agriculture does not aim for a romantic (and unviable) return to pre-industrial forms of production. What it seeks is to set in motion a strategy to modernize farming on the basis of an adequate management of nature and the recognition (rather than destruction) of the rural heritage … What is most surprising (and encouraging) is that this proposal, which has not formed part of either official policies or the debate among local experts, is taking place as a result of commercial transactions, the result of connecting the growing demand for new organic products in the first world with the ecologically oriented production of traditional Mexican communities. This is the case of some indigenous organizations in Oaxaca and Chiapas that have started to supply organic coffee to the … markets in Germany, Italy, Denmark, Holland and other industrial countries. This is because their traditional systems (shade-grown coffee, in mixed farming systems that do not use agrochemicals) managed to survive the policy of modernization. A further example is the consortium of more than a dozen Chinanteca communities [which] have managed to cultivate vanilla … or the producers (and exporters) of sesame …

There are worthy efforts to organize alternative channels of international trade in support of self-managing groups practising ecological agriculture. Would it be possible to commercialize, at higher prices, Andean ecological quinua in Berlin or San Francisco? Let's hope it can be done, but one may doubt (as would Toledo) whether differentiating products into specialized expensive markets is really the most effective method of defending ecological agriculture. Furthermore, in cases like vanilla production competition already comes from the new biotechnology industries.

The problem arises when ecological agriculture cannot compete in the wider market against the products of modern agriculture. When an insoluble conflict arises between ecology and chrematistics – as is generally the case when ecological agriculture and modern agriculture clash – a second option arises. Which social agents will make an ecological economy their political cause? Peasants in the South, who still practice ecological agriculture, seem to be the obvious candidate.

Peasants have always had good ecological practice, as shown by their conservation and creation of genetic resources, their soil and water management systems and their use of renewable solar energy – a source that cannot be taken away, unless they are evicted from their land. However, in the long history of political ideologies based on peasant resistance and their accompanying economic ideas (for example, in Chayanov), there have been no explicitly ecological elements until very recently. Traditional peasants, if they have rights to the land, also have access to the sun's energy and to the rainwater that falls on their land. They also control a 'fourth resource', the seeds from their harvest. Unlike peasants, modern farmers depend on external energy from fossil fuels, they pollute more and they have lost control over this 'fourth resource'.[11] In rich countries, the spread of the market has led to large but overlooked losses of genetic resources. While the countries of the North are poor in genetic resources, some poor countries in the South contain not only the current genetic wealth of tropical rainforests, but are also centres of diversity of many agricultural plants. Furthermore, these countries still have traditional farmers using mixed farming systems with low external inputs, capable of creating and maintaining agricultural diversity and making the most of their own (or their neighbours') genetic resources, as well as of wild varieties.

A relatively new international movement formed by NGOs is struggling to provide an ideological defence for ecological agriculture and those who practice it; it also is spreading new skills and experience independently of governments and commercial companies. Once a word as innocent as 'compost' has spread internationally and forms part of peasants' everyday language, it may become a mental defence against salesmen from fertilizer factories. Once traditional practices receive the internationally recognized name of 'integrated pest management', they acquire a new legitimacy.

7   Economics of ecological agriculture

If only ecological agriculture was chrematistically profitable! Although there are cases in which it is, I think that these are the exceptions to the rule. If these cases were the norm, the commercial sector would be actively present in the production and sale of ecological agricultural products. In fact, this sector is almost exclusively limited to NGOs and traditional farmers from the South. But those who wish to encourage ecological agriculture have a further argument. If only an adequate accounting of externalities – correcting prices and removing subsidies to agricultural chemicals, commercial seeds, and mechanization – managed to conserve or impose ecological agriculture! I believe that the discussion should develop along more directly political lines. Even if we knew how to translate modern agriculture's negative externalities (genetic impoverishment, energy wastage and chemical contamination) into updated chrematistic values (which we don't), we would be prudent to avoid them as much as possible.

Once it has been decided that it is necessary to protect and encourage ecological agriculture, once the question has been argued from the point of view of long-term ecological economics (taking into account uncertainties and the irreversible nature of some events) and once it has obtained enough political force on other groups (e.g., defence of ethnic culture), only then will we be able to calculate the cost – in money, in resources and in hours of labour – of protecting and encouraging ecological agriculture. Nobody claims that this will always be profitable in the short term; it has a cost that it is perhaps worth paying, even though there is no guarantee that it will be recovered in the marketplace.[12]

What costs are we willing to pay for ecological agriculture? What benefits will we obtain? At the moment the only thing that is paid for is the ex situ conservation system. But in situ conservation (and coevolution) has been paid for by the traditional peasantry. We are dealing with a typical case comparing short-term costs (lower apparent production per acre and per hour of labour, for example) and uncertain and diverse benefits in the long-term (the creation and conservation of biological diversity in situ, lower contamination, savings in fossil fuels). Conventional economics does not help us when it comes to making this decision. Thus, the international movement in favour of ecological agriculture should not worry about short-term economic chrematistic considerations. It should be a political movement that appeals to ecological-economic reasoning (and to other lines of reasoning, such as the defence of peoples whose ethnic identity and farming systems are threatened). It should not discuss biological diversity in terms of the monetary value of its immediate use (or at least not only in these terms) nor in terms of a hypothetical future usefulness (which has been termed their option value), but, above all, in terms of an existence value that can hardly be reduced to money. This will link the struggle for ecological agriculture with the wider struggle for the conservation of wild biological diversity.

8   The NAFTA: petroleum and corn

To conclude, let us consider how the defence of agricultural diversity might be raised in Mexico today, in the light of what has been said so far. In the US, petroleum is relatively cheap, although the country is now one of the major petroleum importers. Mexico exports cheap petroleum to the US. It is 'cheap' because it does not take into account the ecological costs of extraction in the Campeche and Tabasco areas nor the costs of carbon dioxide (and NO_x) emissions. In addition, the price implicitly undervalues Mexico's future demand for petroleum. Putting an ecological tax on petroleum in Mexico would lead to a conflict. As things are now, within the NAFTA framework, Mexico cannot levy taxes upon its exports. Thus, it will export cheap petroleum to the US and it will import products (such as corn) that are in part produced using cheap Mexican petroleum.

This American corn is of little genetic interest and (in part) requires a continuous supply of Mexican genetic resources, which have so far been free. US corn exports are subsidized and will continue to be, at least to the extent that their price does not include any accounting entry corresponding to ecological costs. These exports will damage peasant corn production in Mexico, which is more efficient in terms of low fossil energy use and also more biologically attractive. In other words, US agriculture has lax environmental standards in comparison with Mexican peasant agriculture. So any assessment of the impact that liberalization of corn imports would have in Mexico, in terms of benefits and costs (Levy and van Wijnbergen, 1991), must include some estimate of environmental costs and benefits.

What would be the environmental costs of the growth in some parts of the Mexican economy due to the NAFTA? Various US ecological groups paid great attention to the export of Mexican tuna to the US, because the tuna was caught by methods leading to the death of dolphins. They have also paid attention to the potential effects of the NAFTA in increasing sweatshop production just over the border, as well as other economic activities, such as strawberry production, which are subject to less strict environmental regulation in Mexico than in the US. These are without doubt important issues, as is the possible export of domestic and industrial (including nuclear) waste from the US to Mexico. But crucial points in the ecological-economic discussion, because of their sheer quantity, should be the environmental costs of cheap petroleum exports from Mexico and the threat to its ecological farming system and food security.

The conclusion of the first negotiating round of the NAFTA in the summer of 1992 was well received by US corn (and pork) producers, who anticipated a big rise in exports to Mexico. It was correctly said that Mexican barriers to the import of corn have prevented US farmers from dominating the Mexican food market and ruining hundreds of thousands of farmers in southern Mexico (e.g., *New York Times*, 15 August 1992, p. 34). Now, as a result of NAFTA, Mexico must immediately allow the tariff-free importation of 2.5 million tons of cereals every year. Further, the tariffs on imports exceeding this figure would be gradually reduced to zero over 15 years. It is said that this free-trade policy would benefit both countries, as US corn is produced more efficiently than Mexican corn. But how can we talk of efficiency without previously agreeing on a measure of agricultural productivity that takes into account fossil fuel use and the loss of biological diversity in modern agriculture? Perhaps the best system would combine the ecological superiority of traditional Mexican Papa agriculture (without doubt excessively based on hard human labour) and the economic-chrematistic superiority of US agriculture (which does not take into account the negative externalities it produces). Ecological criticism of conventional agricultural economics leaves plenty of room for different political points of view because, while this critique shows that the current prices are incorrect, it is unable to say what the ecologically correct prices should be in order to fully internalize new externalities.

One cannot condemn this ecological critique as if it were an excuse for obstinate nationalist protectionism, nor, in my opinion, does it make sense to defend the idea of 'bioregional' units totally closed to foreign products (and citizens) from an ecological perspective. On the contrary, from such a perspective one ought to argue that horizontal transport of elements present in abundance in one territory, but limiting factors (in Liebig's sense) in another, will increase the joint capacity to support life. Of course, horizontal transport is not without costs and an adequate accounting system would include its ecological costs. But an argument for trade based on Liebig's law is different from an argument for trade based on absolute comparative chrematistic advantages.[13]

Ecological opposition to the NAFTA should insist above all on the fact that no charge is made for environmental costs in export prices, whether petroleum from Mexico or corn from the US. But this does not mean that we have discovered a magic method of establishing the 'total environmental costs' of those economic activities with future ecological consequences that are irreversible and unknown. It needs to be stressed that there are no ecologically correct prices in the sense that they convincingly internalize all externalities. Even so, prices, may be *ecologically corrected* to take environmental externalities into account, e.g., by putting a tax on Mexican petroleum in Mexico and by putting taxes in the US on agricultural production using modern technology. Trade flows would then be based on comparative or absolute advantages that have been ecologically corrected. But these taxes would be in opposition to the free trade ideology expressed in the NAFTA and do not figure in the US political agenda, because of their distributive effects. The US may consider an ecological tax on Mexican petroleum, but to be paid in the US rather than in Mexico and awareness of the fact that US agriculture uses 'dirty' technologies with negative environmental impacts is still not widespread. These taxes may well become a subject for political discussion in Mexico, where there have been traditions of political 'agrarismo' since the times of Zapata and of petroleum nationalism since Cardenas in the 1930s, both of which could easily be connected with the new ecological awareness.[14]

Notes

1   It is difficult to construct indicators of genetic erosion. The names of varieties used in traditional agriculture have often not been recorded and the extent of farmers' utilization of seeds is unknown. Which varieties occupy a given share of a crop within a given genetic distance is difficult to determine from their popular names.

2   Personal communication from Daniel Querol.

3   French acronym for the Union for the Protection of New Plant Varieties.

4   This is the case even in the European Union, where, besides a program for animal species, there are now also proposals for in situ conservation of plant genetic resources.

5   A well known Peruvian example: in 1638, the countess of Chinchon, the Viceroy's wife, became ill with malaria. She recovered by taking a medicine made out of the quinine tree by her indigenous servant (whose name is not recorded). For further stories on quinine, see Brockway, 1988.

6   Cf. Nanjundaswamy, 1993; Sahai, 1993. For Latin America, see Montecinos, 1993; Tapia, 1993 and for several contributions on Mexico, see Leff and Carabias, 1993.
    CGIAR is a consulting group of institutes of agronomic research, partly controlled by the World Bank.

7   Not yet ratified by the US Senate.

8   Some fervent neo-liberals propose that the way to save the whales is by privatizing them. Thus, if the members of Greenpeace and other groups pay more than Japanese or Norwegian fishermen, saving the whales will be achieved through the market. In the same way, entomological societies would have to pay to prevent the disappearance of insect species. However, in these marketplaces future generations are absent and, of course, the threatened species or varieties in question cannot bid.

9   Herman Daly's classification includes natural capital, human-made capital and, as a special case, cultivated natural capital. Daly has raised

the question (already raised by Frederick Soddy) of whether these categories of capital substitute for each other or are complementary.

10 For instance, Dr Patarroyo of Colombia seems to have obtained a malaria vaccine chemically. He will gain many honours and prizes for his successful work, the patent for which he gave away to the World Health Organization, without royalties.

11 This is the terminology used by Henk Hobbelink, the founder of GRAIN (Genetic Resources Action International), a NGO based in Barcelona, which supplies information about the threats to and importance of agricultural diversity (c/o Girona 25, Barcelona 08010).

12 A further example: the defence of ecosystems like mangrove swamps in Ecuador and other countries where they are commercially threatened by ponds for raising shrimps, which are also used by part of the local population as a source of firewood and seafood. Mangrove swamps also fulfil important ecological functions. There is no guarantee that chrematistic advantage is on the side of conservation.

13 Ricardo did not actually use money but labour value as a common measure. My argument against free trade is based on the existence of different standards of value, i.e., on economic incommensurability.

14 This article is based on a paper given at the International Conference on Biological Diversity in Ibero-America, CIELAT, University of the Andes, Merida, Venezuela, 19–24 October, 1993. A revised version of the paper was given at EAERE Conference, Dublin, 22–24 June 1994. This version is reprinted with revisions from *Etllnoecologia* 3, 1994. 1 am grateful to Cristina Marco and Trevor Foskett for comments, and to Patricia Allen for editorial help.

References

Brockway, L. (1988), 'Plant Science and Colonial Expansion: The Botanical Chess Game' in J. Kloppenburg (ed.), *Seeds and Sovereignty. Tile Use and Control of Plant Genetic Resources*, Duke University Press: Durham and London.

Cooper, D., Vellvé, R. and Hobbelink, H. (eds) (1992), *Growing Diversity: Genetic Resources and Local Food Security*, Intermediate Technology Publications: London.

Descola, P. (1988), *La Selva Culta. Simbolismo y Praxis en la Ecologia de los Achuar*, Abya Yala: Quito.

FAO (1993), *La diversidad de la naturaleza: an patrimonio valioso*, Rome.

Gámez, R. (1992) in E.A. Brugger and E. Lizano (eds), *Eco-eficiencia. La vision empresarial para el desarrollo sostenible en América Latina*, Oveja Negra, Business Council for Sustainable Development: Bogota.

Grilliches, Z. (1958), 'Research Cost and Social Returns: Hybrid Corn and Related Innovations', *Journal of Political Economy*, No. 66.

Guha, R. and Gadgil, M. (1992), *This Fissured Land. An Ecological History of India*, Oxford University Press: Delhi.

Leff, E. and Carabias, J. (1993) (eds), *Cultura y manejo sustentable de los recursos naturales*, M.A. Porrua y UNAM: Mexico.

Levy, S. and van Wijnbergen, S. (1991), 'El maiz y el acuerdo de fibre comercio entre Mexico y los Estados Unidos', *Ed Trimestre Economico*, No. 131.

McNeely, J.A., Miller, K.R., Reid, W.V., Mittermeir, R.A. and Werner, T.B. (1990), *Conserving the World's Biological Diversity*, IUCN, WRI, cl, WWF-US, World Bank: Gland [Switzerland] and Washington DC.

Martinez-Alier, J. (1990), *Ecological Economics*, Blackwell: Oxford and New York.

Montecinos, C. (1993), 'Las negociaciones internacionales sobre recursos geneticos', *Ecología Política*, No. 5.

Nanjundaswamy, M.D. (1993), 'Farmers and Dunkel Draft', letter to the editor in *Economic and Political Weekly*, 26 June.

Posey, D. (1985), 'Indigenous management of tropical forest ecosystems: the case of the Kayapo Indians of the Brazilian Amazon', *Agroforestry Systems*, Vol. 3, No. 2.

Querol, D. (1987), *Recursos geneticos nuestro tesoro olvidad*, Industrial Clrafica: Lima.

Querol, D. (1992), *Genetic Resources: Our Forgotten Treasure*, Third World Network: Penang.

Reid et al., W.V. (1993), *Biodiversity Prospecting: Using Cenetic Resources for Sustainable Development*, World Resources Institute: Washington DC.

Reyes, V. (1996), 'The Value of Sangre de Drago Seedling' in the quarterly newsletter of GRAIN, Barcelona, March 1996, Vol. 13, No. 1.

Richards, P. (1984), *Indigenous Agricultural Revolutions: Ecology and Food Production in West Africa*, Hutchinson: London.

Rocheleau, D. (1991), 'Gender, Ecology and the Science of Survival: Stories and Lessons from Kenya', *Agriculture and Human Values*, Winter–Spring.

Sahai, S. (1993), 'Dunkel Draft is Bad for Agriculture', *Economic and Political Weekly*, 19 June.

Tapia, M. (1993), 'Gestion de la biodiversidad andina' in B. Marticorena (ed.), *Recursos Naturales: Teceologia y Desarrollo*, Centro 'Bartolome de Las Casas': Cusco.

Toledo, V.M. (1988), 'La sociedad rural, los campesinos y la cuestión ecológica' in J. Zepeda (ed.), *Las Sociedades Rurales Hoy*, El Colegio de Michoacán: Conacyt.

Toledo, V.M. (1989), 'The Ecological Rationality of Peasant Production' in M. Altieri and S. Hecht (eds), *Agroecology and Small Farm Development*, CRC Press: Boca Ratón, Florida.

Toledo, V.M. (1991), 'La sociedad rural, los campesinos y la cuestión ecológica', *Ecología Política*, No. 1.

Toledo, V.M. (1992), 'Regresemos al Agro', *Cuadernos Verdes*, Colegio Verde de Villa de Leyva, 5.

Vellvé, R. (1992), *Saving the Seed. Genetic Diversity and European Agriculture*, Earthscan: London.

World Resources Institute (1992), *World Resources 1992–1993*, Oxford University Press: New York.

Contributors

Robin Attfield is Professor of Philosophy at the University of Wales, Cardiff.

Ted Benton is Professor of Sociology at the University of Essex.

David Cooper is Professor of Philosophy at the University of Durham.

Tim Gray is Professor of Political Thought at the Department of Politics, University of Newcastle.

Tim Hayward is Senior Lecturer in the Department of Politics, University of Edinburgh.

Chris Himsworth is Senior Lecturer in the Department of Public Law, University of Edinburgh.

Russell Keat is Professor of Political Theory, Department of Politics, University of Edinburgh.

Donald McGillivray is Lecturer in Environmental Law, Kent Law School, University of Kent at Canterbury.

Joan Martinez-Alier is Professor of Economics and Economic History, Universidad Autonoma, Barcelona.

Markku Oksanen is Research Fellow in the Department of Philosophy, University of Turku, Finland.

John O'Neill is Reader in Philosophy, Department of Philosophy, Lancaster University.

Kate Soper is Senior Lecturer in Philosophy at the University of North London.

Anthony Stenson is a PhD student in the Department of Politics, University of Newcastle.

John Wightman is Lecturer in Law, Kent Law School, University of Kent at Canterbury.